Kuehne Foundation Book Series on Logistics

Volume 13

Haupt

Stephan M. Wagner
Christian Busse
Editors

Managing Innovation

The New Competitive Edge for Logistics Service Providers

Haupt Verlag
Berne · Stuttgart · Vienna

Stephan M. Wagner is Professor of Business Administration, holds the Kuehne Foundation Endowed Chair of Logistics Management and is Director of the Kuehne-Center for Logistics Management at WHU – Otto Beisheim School of Management, Vallendar, Germany. Prior to joining academia, he was director of supply chain management for a Swiss-based technology group and senior manager at a renowned consulting firm. He is author and editor of eight books and nearly hundred book chapters and articles.

Christian Busse is Ph.D. student and research assistant at WHU – Otto Beisheim School of Management, Vallendar, Germany. He studied Industrial Engineering at the Technical University of Berlin and the Universidad Complutense de Madrid. He worked as market researcher in London and as consultant for The Boston Consulting Group where he focused on strategic supply chain projects and the management of LSPs.

1st edition: 2008

Bibliographic information published by *Die Deutsche Nationalbibliothek*
Die Deutsche Nationalbibliothek lists this publication in the Deutsche Nationalbibliografie; detailed bibliographic data are available in the internet at http://dnb.d-nb.de

ISBN 978-3-258-07340-8

Printed in Switzerland

www.haupt.ch

Preface

The contributions presented in this book range across a number of disciplines, including logistics, management, innovation, and marketing. They are united, however, by a theoretical core: the perception that innovation management at logistics service providers (LSPs) is vital to the competitiveness of firms in the industry. Since managing innovation in the logistics service industry is so multi-faceted and this book is the first in this area, some fundamental questions will be addressed. This book presents and provides solutions for this timely and challenging topic, and can serve as a guideline for managers in the logistics service industry. Everyone with an interest in innovation and logistics will find this book valuable and thought-provoking.

We are thankful to a large number of people who contributed to the success of our book project. First and foremost are all of the contributing authors, each of whom is a thought leader from academia or corporate practice. Second, we thank the sponsor of this publication, the Kuehne Foundation, and in particular Klaus-Michael Kühne, its president. Third, Martin Willhaus, from the Kuehne Foundation encouraged us to start the adventure of publishing a book in the Kuehne Foundation Book Series on Logistics, and Elisabeth Homberger from the Haupt publishing company who supported us in compiling and promoting the book.

We received help from numerous colleagues at WHU – Otto Beisheim School of Management throughout the project. In particular, we would like to thank Andreas Bahke, Ulrike Cappel, Eckhard Lindemann, Jessica Quiroz, Carl Marcus Wallenburg and Jürgen Weber.

Finally, we are grateful to our partners, Nicole Wagner und Ulrike Volz, for their patience over the course of this project and their understanding that it took away a substantial amount of our time.

Stephan M. Wagner and Christian Busse — Vallendar, January 2008

Table of Contents

Chapter 1

Managing Innovation at Logistics Service Providers – An Introduction

Stephan M. Wagner and Christian Busse

Abstract

This chapter is a prologue to the management of innovation by logistics service providers (LSPs). Both innovation and its management are defined and environmental trends influencing LSPs are discussed. A brief review of milestones in the exploration of LSPs' innovation management is provided. The chapter closes with an overview of the other contributions in the book.

1 Managing Innovation at Logistics Service Providers – An Introduction

1.1 Relevance

Firms in the logistics service industry have to operate and survive in a competitive marketplace characterized by, for example, legislative changes, deregulation and consolidation, and shippers' demands for ever-higher levels of service. The competitiveness of logistics service providers (LSPs) increasingly depends on their ability to adopt innovations that add value to shippers' bottom line. Likewise, innovative LSPs can reap first-mover advantages, increase customer satisfaction and retention, and capture larger shares of the market. However, empirical evidence suggests that today the capacity of LSPs to generate innovation is rather low (Wagner, 2007). They turn out fewer service and process innovations than firms in other industries (Wagner, 2008). Since "logistics research has largely ignored the concept of innovation" (Flint et al., 2005, p. 113), it is worthwhile to take a closer look at the industry.

The investigation of innovation management at LSPs is relevant both to academics and to forward-looking practitioners. The book is intended for both of these groups. It helps managers to better understand how LSPs can more actively approach innovation management, how they can generate and manage innovation at the shipper-LSP interface, and how they can make innovation happen. In addition, the book establishes a solid foundation for a new and fruitful research stream.

1.2 Innovation and Its Management

There are plenty of ways to capture the notion of "innovation" (Garcia and Calantone, 2002). For this book, we define innovation in the following manner: Innovation is a subjective novelty which is the result of a conscious management process and which aims at economic exploitation. We want to elaborate briefly on the four essential constituents.

Novelty

As the term suggests, innovation implies novelty. The degree of novelty, however, is not predetermined. Hence, it can be incremental, radical, or anywhere in between. As no restriction shall be made concerning the object of innovation, the term can be applied to a novelty in what is being sold (traditionally labeled "product innovation" or "service innovation") as well as to novelty in processes ("process innovation").

For LSPs, it is particularly appropriate to include both product and process innovations in the investigation of innovation. This decision is founded first on the immateriality of logistics services which complicates the distinction between product and accompanying processes. Second, it is founded on the complexity of terms: the term "product" is often reserved for standardized implementations of services, whereas individual and customized implementations are often referred to as "solutions." It is also appropriate not to discriminate against relatively low levels of novelty, as many logistical innovations cannot reach radical novelty.

Subjectivity

As the difficulty of introducing something new into a company arises from its novelty to the company, "subjective" novelty shall be regarded as sufficient novelty to apply the term innovation: "An innovation is [...] new to an organization and to the relevant environment" (Knight, 1967, p. 478). Of course, a global novelty is still likely to be more difficult to introduce than a merely subjective (local) one, in part due to the diffusion of knowledge.

As a consequence of this definition, LSPs can and must regard imitations (e.g., local, but not global novelties) as innovations. This makes additional sense from a management point of view, as the decision to "make or copy" can thus be taken by the same actor who is responsible for the overall management of an innovation.

Manageability

If there was just unmanageable "random innovation," a discussion of the phenomenon under observation would be superfluous. A similar result might stem from somewhat innovative efforts which at best can only result in very limited improvements. It might be rational not to apply a conscious management process to those. All other innovation efforts have to be managed.

Innovation processes might be more complex, incalculable and risky than routine decisions or processes. In addition, friction and conflict can often be observed within innovation projects. Hence, innovation management differs from the management of routine decision-making because it requires a large amount of flexibility and is less open to detailed preliminary planning.

Exploitation Purpose

Schumpeter (1936) was the first to differentiate between an invention as a technical novelty with no relevance to the market, and an innovation which includes a novelty's exploitation. For that reason, at least in the context of business research, the term "invention" might be reserved for novelties which prove to be successful within a market or within an intra-firm application. There is one problem, however: Given frequently occurring up-front costs (such as concept development costs, market introduction costs etc.), the actual economic success of a supposed innovation can only be judged at the end of its lifetime. It is appropriate not to exclude developments which are supposed to pay off economically, but do not. In order to prevent neglecting economically less successful inventions, the only distinct attribute concerning the commercial success is the intention to achieve it.

The above argument results in a sequence of outcomes which are summarized and depicted in Figure 1.

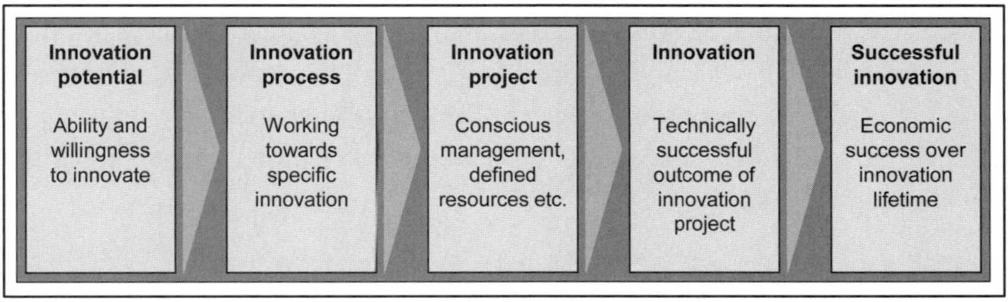

Figure 1: Sequence of Outcomes

Innovation Management

Applying the manageability criterion to innovation management, we adopt a comprehensive scope regarding our subject (Figure 1). Beyond the management of individual innovation processes or projects we subsume under innovation management also the design of the innovation management system within a company or business unit. That necessarily includes the definition of an innovation strategy and its alignment with that company's strategy as well as the development of organizational structures which are adequate to encourage innovation, and to control particular innovation processes. Furthermore, it should encompass the imprint of an innovation culture, include the development of appropriate support instruments for innovation processes, and permit the management of particular innovation processes and their interdependencies. All of these components will be referred to periodically throughout the book.

1.3 Innovation-relevant Features of Logistics Service Providers

The LSP industry seems quite heterogeneous at first sight. There are carriers, forwarders, courier, express and parcel (CEP) service providers, third-party logistics providers (3PLs) and possibly others (e.g., fourth-party logistics providers). The international literature deals mostly with 3PLs (e.g., Hertz and Alfredsson, 2003; Langley et al., 2006; Lieb, 2005).

There is some danger that the – admittedly existing – differences among various business models and service bundles of LSPs might obscure their commonalities. Since substantial effort has been invested in the study of service innovation in general (i.e., independent of a particular industry), the consequential next step is to investigate innovation management for the logistics industry on aggregate (i.e., for all LSPs) before studying innovation management for a sub-set of the logistics service industry.

Environmental Factors

One of the challenges LSPs face is a pronounced globalization of their customer industries: US-American high-tech companies employ East Asian contract manufacturers as subcontractors, originally German automotive OEMs make use of production facilities and supplier parks in South Africa, and Indian steel giants buy numerous Western plants and even competitors. This trend is facilitated by mobility and availability of production factors. Simultaneously, LSPs want to make use of effects of scale and of network effects. Accordingly, there is a trend towards the globalization of logistics markets which is accompanied by a consolidation trend. Fierce global competition raises the requirements LSPs have to fulfill and the challenge to grow profitable (Chakravarthy and Lorange, 2008). Innovation can be an answer to this.

At least since the 1990s companies have begun to concentrate on their core competencies. Consequently, a lot of outsourcing has been visible; most of all in transport and warehousing. LSPs have broadened their service offerings (Lieb, 2005) to allow outsourcing of more complex services such as value-added services. This potential can likewise be exploited by means of innovation.

Frequent Characteristics

Given their origin in relatively simple transport and warehousing services it is hardly surprising that LSPs' employees have their feet firmly planted on the ground. This might explain a lack of innovation tradition within the industry and a rather reactive approach to innovation (e.g., Wagner, 2007).

The traditionally high percentage of simple operational tasks means that across LSPs the employees on average tend to have a relatively low level of education and formalized training.

Most LSPs are markedly decentralized organizations with distributed operations and sales activities. This is consistent with a tendency to decentralized decision making and to decentralized P&L responsibilities (e.g., through profit centers).

Individual customers quite often have a tremendous importance for LSPs. Their processes are frequently interlocked with their customers' processes, they have verticalized service offerings for individual industries (Lieb, 2005) and their business often has a project character.

Finally, LSPs offer services to their customers. Services are immaterial and intangible. In comparison with physical products, services are abstract, more difficult to plan and to test. Intellectual property concerned with services is more difficult to protect. Services cannot be stored because they are consumed while they are produced. Logistics services in particular are rendered around the physical goods of LSPs' customers.

The innovation-relevant features of LSPs' environment emphasize the need to develop innovations and therefore a suitable innovation management. The common characteristics of LSPs determine its design. We will refer to these characteristics throughout the book.

1.4 A Brief Literature Review

Most articles linking "innovation management" and "logistics" were published very recently; only one article predates 2000 (Meyers and Tucker, 1989). In total, 28 publications written in English were identified and grouped into four research streams. One of those concerns innovation (management) at LSPs. The other three are auxiliary to our effort (Figure 2).

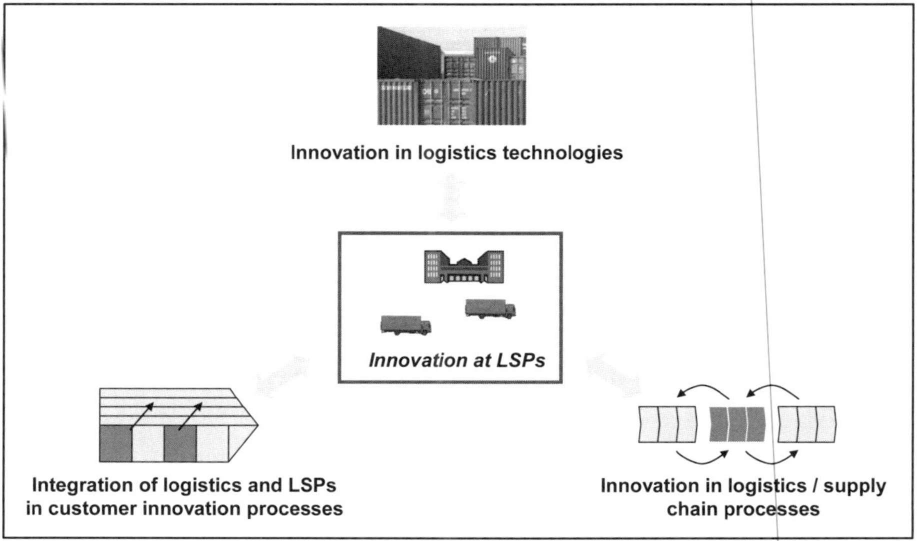

Figure 2: Research Streams Linking Innovation Management and Logistics

Below we outline the most important publications touching upon innovation at LSPs.

Meyers and Tucker (1989) investigate the integration of (corporate) logistics into new product development (NPD) to produce technological innovations. Comparing routine innovations and radical innovations, logistics integration seems to be particularly helpful for incremental innovations. Zacharia and Mentzer (2007) follow this line of thought. By means of a large-scale empirical survey they could identify benefits of logistics integration into NPD.

Several papers have broached the issue of LSP innovation management based on the cases of ten distribution centers in Australia and Singapore: Hyland et al. (2003, p. 334) investigate individual and collective learning behavior (with a focus on supply chains). Concerning the LSPs they state "[t]here was no clear indication one way or the other that the firms displayed the ability to actively manage knowledge." Soosay and Hyland (2004) analyze innovation drivers and classify them as external and internal and into push and pull. Soosay (2005) hints at the individual actors' abilities for continuous innovation at an LSP. To this, Soosay and Chapman (2006) add an analysis of performance indicators. Soosay and Hyland (2005) investigate contingency factors. They find that the ownership structure is meaningless for the innovative capacity of firms, and that larger firms are better suited to carry innovation process risks than smaller ones. Finally, Soosay and Sloan (2005) use the same ten cases to investigate change processes within LSPs. Customer satisfaction is the most important driver of change in the case of Australian distribution centers, whereas in Singapore intended continuous improvement is eminent.

Flint et al. (2005) engage in their explorative work with "logistics innovation." Logistics innovations is found to be the result of a four-phase process: "setting the stage," "customer clue gathering," "negotiating, clarifying, and reflecting" and "inter-organisational learning." In addition, their article stresses the meaning of customer orientation and integration in all innovation processes.

Cahill (2007) presents the results of a large-scale empirical study of customer retention in the logistics sector. He shows that an LSP striving for improvements will experience a steep increase in customer loyalty. These findings are confirmed and extended by Deepen (2007) in respect to the positive influence which striving for improvements may have on the logistics outsourcing performance of customers.

1.5 An Overview of the Individual Contributions

Another nine chapters in this book, written by experts from Europe, North America and Asia, will shed light on the management of innovation for LSPs. While their points of view and their focuses vary, they complement one another.

Before an LSP can produce an innovation, it has to overcome many barriers, a number of which found throughout the LSP industry. *Oke* analyzes these barriers, thus making them addressable for an LSP's management. He therefore makes a very significant contribution to the establishment of an LSP-focused research stream.

Speh amends a chapter on the marketing of innovative logistics services. In so doing, he leaves the usual line of thought of customer-specific on-demand developments behind and allows LSPs to adopt a more proactive position in their innovation management.

De Souza and *Goh* depict a strategy model which an LSP can apply in order to stay ahead of its competition. They accompany it with a case example, the YCH group, an Asian LSP, thus adding a regional perspective to the book.

Flowers, *Cort*, *Artman* and *Ballou* establish the vital link between strategy and processes. In the first step, they demonstrate how open innovation is compatible with stage-gate processes, a result which is meaningful even beyond the LSP world. Secondly, they tie the resulting process model to three market leader models for LSPs. Their finding here, namely that LSPs can successfully apply all three market leader models, is meaningful beyond the innovation management context of this book.

An LSP-specific perspective on innovation processes is provided by *Busse and Wagner* who devise an audit tool allowing managers, consultants or academics to analyze an LSP's innovation processes. It includes descriptions of IT tools and management instruments to be used as well as phase-specific key success factors.

Gammelgaard adds a theoretical supply chain perspective to innovation processes. Following the case example of Maersk Logistics, she emphasizes the importance of customer interaction, discloses four pitfalls awaiting LSPs, and describes ways of avoiding them.

Franklin supplements a management perspective to innovation processes. Using Kuehne + Nagel as a case of successful implementation, he applies the concept of "democratized innovation" to channel an LSP's innovation efforts based on management-developed platforms and toolkits.

The customer interaction line is further followed by *Zacharia*. He discusses the opportunities of LSP involvement into their industrial customers' new product development. His contribution is not only theoretically well-founded, it also opens up new horizons as he significantly extends the range of classical LSP services towards R&D consulting.

Busse and *Wagner* close the book with a generic implementation guideline which should help LSPs in the implementation of an innovation management system.

References

Cahill, D.L. (2007). *Customer Loyalty in Third Party Logistics Relationships – Findings from Studies in Germany and the USA*, Berlin: Springer.

Chakravarthy, B; Lorange, P. (2008). *Profit or growth? Why you don't have to choose*, Upper Saddle River, NJ: Wharton School Publishing

Deepen, J.M. (2007). *Logistics Outsourcing Relationships – Measurement, Antecedents, and Effects of Logistics Outsourcing Performance*, Berlin: Springer.

Flint, D.J.; Larsson, E.; Gammelgaard, B.; Mentzer, J.T. (2005). Logistics Innovation: A Customer Value-Oriented Social Process, *Journal of Business Logistics*, 26 (1), 113-147.

Garcia, R.; Calantone, R. (2002). A Critical Look at Technological Innovation Typology and Innovativeness Terminology: A Literature Review, *Journal of Product Innovation Management*, 19 (2), 110-132.

Hertz, S.; Alfredsson, M. (2003). Strategic Development of Third Party Logistics Providers, *Industrial Marketing Management*, 32 (2), 139-149.

Hyland, P.W.; Soosay, C.A.; Sloan, T.R. (2003). Continuous improvement and learning in the supply chain, *International Journal of Physical Distribution & Logistics Management*, 33 (4), 316-336.

Knight, K.E. (1967). A Descriptive Model of the Intra-Firm Innovation Process, *Journal of Business*, 40 (4), 478-496.

Langley Jr., C.J.; van Dort, E.; Ross, T.; Topp, U.; Allen, G.R.; Sykes, S.R. (2006). *2006 Third-Party Logistics, Results and Findings of the 11th Annual Study*, Atlanta, GA: Georgia Institute of Technology.

Lieb, R.C. (2005). The 3PL Industry: Where it's Been, Where it's Going, *Supply Chain Management Review*, 9 (6), 20-27.

Meyers, P.W.; Tucker, F.G. (1989). Defining Roles for Logistics During Routine and Radical Technological Innovation, *Journal of the Academy of Marketing Science*, 17 (1), 73-82.

Schumpeter, J. (1936). *The Theory of Economic Development*, Cambridge, MA.

Soosay, C.A. (2005). An Empirical Study of Individual Competencies in Distribution Centres to Enable Continuous Innovation, *Creativity and Innovation Management*, 14 (3), 299-310.

Soosay, C.A.; Chapman, R.L. (2006). An Empirical Examination of Performance Measurement for Managing Continuous Innovation in Logistics, *Knowledge & Process Management*, 13 (3), 192-205.

Soosay, C.A.; Hyland, P.W. (2004). Driving Innovation in Logistics: Case Studies in Distribution Centres, *Creativity and Innovation Management*, 13 (1), 41-51.

Soosay, C.A.; Hyland, P.W. (2005). Effect of Firm Contingencies on Continuous Innovation, *International Journal of Innovation and Technology Management*, 2 (2), 153-169.

Soosay, C.A.; Sloan, T.R. (2005). Driving Change: Innovative Management in Distribution Centres, *Journal of Asia Entrepreneurship and Sustainability*, 1 (2) 1-21.

Wagner, S.M. (2007): Logistics Industry: The Future Belongs to Innovative Logisticians, *International Transport Journal*, 69 (35/36), 14-15.

Wagner, S.M. (2008): Innovation Management in the German Transportation Industry, *Journal of Business Logistics*, 29 (2), forthcoming.

Zacharia, Z.G.; Mentzer, J.T. (2007). The Role of Logistics in New Product Development, *Journal of Business Logistics*, 28 (1), 83-110.

Chapter 2

Barriers to Innovation Management in Logistics Service Providers

Adegoke Oke

Abstract

The manner in which innovation is managed in service companies has been of great interest because of the challenges associated with the nature of services as opposed to tangible manufactured products. For logistics services providers (LSPs), innovation management presents additional challenges. In this paper, the factors that are likely to inhibit innovation management in LSPs are identified and discussed. The identification of these barriers constitutes a first step in improving innovation management in LSPs.

2 Barriers to Innovation Management in Logistics Service Providers

2.1 Introduction

The impact of the service sector on the economies of western countries is on the rise. With the increasing outsourcing of traditional manufacturing operations to low-cost economies of Asia, there has been a growth in the service sector contribution to the GNP of major western economies. For example, in 2005, service was responsible for 50% of GNP in the United Kingdom, an increase of about 15% over the previous 10 years (Economist Intelligence Unit, 2006). The service sector is comprised of the transport, government, education, health care, social and personal services, retail and wholesale, hotels and restaurants, telecommunications, and financial services sectors. Logistics, a major component of the transport sector, has been identified as a major service sector in developed economies. For example, in the US, transportation and warehousing account for over 10% of GDP, amounting to about $900 billion in 1998 (Delaney, 1999). In spite of the growing importance of the service sector, research in this sector has lagged behind studies in the manufacturing sector (Griffin, 1997; Oke, 2004). In particular, studies on innovation management have been biased towards the manufacturing sector. This is not surprising given that innovations have traditionally been associated with tangible manufactured products. In addition, the intangible nature of services makes them less amenable to innovation management studies.

Within the service sector itself, less attention has been paid to logistics service providers (LSPs). Logistics consists of all activities involved in the supply of a service or a product to a customer. An LSP is a company whose primary activity is the provision of logistics services, including warehousing and transportation. With the global trend in outsourcing, LSPs play an important role in global supply chain systems. However, competition by LSPs to win businesses is fierce. Manufacturers want to partner with LSPs that can ensure speed, cost-effectiveness and reliability. In a globalizing market, more and more manufacturers need help in managing complex supply chains. Original equipment manufacturers (OEMs) favor LSPs who have potential for cross-selling and highly innovative solutions and at the same time can collaborate in planning and executing their logistics strategies. To meet these customer demands, LSPs need to be innovative in their offerings and in how they do business. While many LSPs are engaging in acquisitions and joint ventures in order to meet their customer needs, the requirement for innovative solutions remains high for LSPs. At the same time, the challenges of making innovations happen in LSPs are onerous.

The objective of this paper is to examine some of the barriers to innovation management in LSPs. The identification of such barriers is a first step in understanding how innovation management practices can be improved in LSPs. This paper is based on the results of a study of 63 senior managers in service companies in the UK and on interviews with senior managers in multinational LSPs in addition to secondary data sources. The survey was based on questionnaire items that sought to identify the barriers to innovation management in the service sector. The questionnaires were mailed to senior managers. In addition, semi-structured interviews were held with senior managers in selected service companies who had responsibilities for managing innovation and new product development. The results of the interviews were combined with the analysis of the survey findings to report the key barriers that service companies face in managing innovation. In order to appreciate these barriers, the generic processes for managing innovations in organizations are examined. Then, a discussion of barriers to innovation management in LSPs is presented leading to implications and recommendations for LSPs wishing to improve their innovation performance.

2.2 Innovation Management in LSPs

While innovation management in LSPs has not attracted much attention in the literature, studies on innovation management in service companies as a whole can provide useful insights. However, these latter streams of studies themselves draw heavily from innovation management studies in the manufacturing sector. For example, many research studies on innovation management in service companies have used frameworks such as Booz, Allen and Hamilton's (1982) framework that was based on tangible manufactured products (Figure 1).

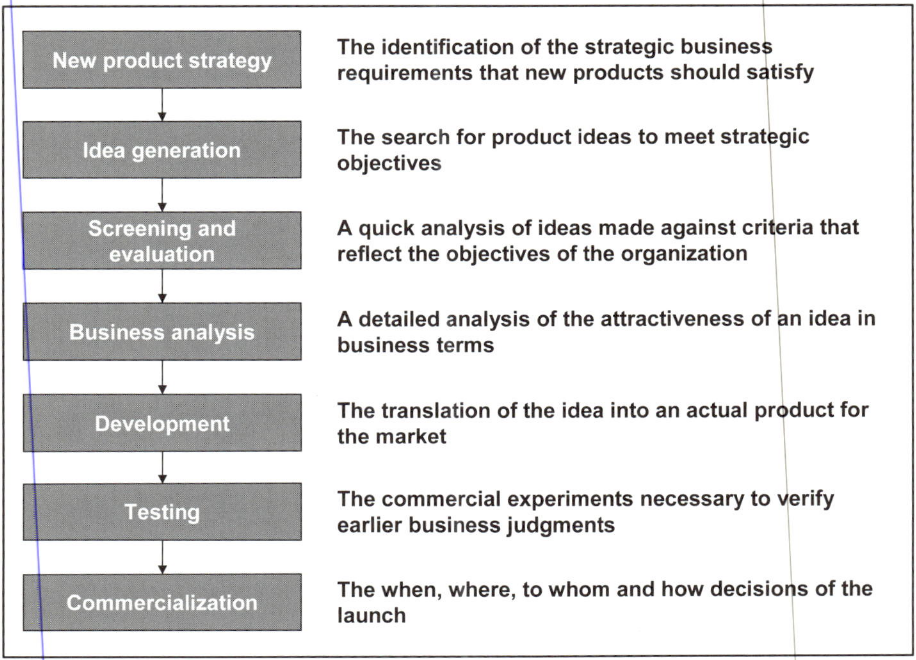

Figure 1: New Product Development Model (Source: Booz, Allen and Hamilton, 1982)

The framework gives an impression of a linear innovation process starting with the definition of an innovation strategy or a new product strategy. In terms of idea generation and concept development within the framework, the focus is on ideas that meet the defined strategic objectives. Subsequent stages in the process include selection and evaluation of ideas and implementation. The 'Pentathlon' framework (originally developed by Goffin and Pfeiffer (1999) from a comparative study of innovation management in German and UK manufacturing companies – Figure 2) has been used to study innovation management practices in service companies (Oke, 2007). The pentathlon framework is similar to the Booz, Allen & Hamilton framework except that the former differentiates process activities (idea generation, selection and implementation) from non-process activities (innovation strategy and human resource management). The principle behind the pentathlon is that good performance in all five areas of innovation management is required to ensure good innovation performance. The innovation strategy provides the direction and focus areas for innovation for an organization while the human resource management aspect can identify management actions and priorities for encouraging innovation in the workforce and creating an environment and a culture where innovation thrives. The middle portion of the framework represents the process of capturing and managing ideas, selecting the best ones to implement and actually managing the implementation process (Figure 2).

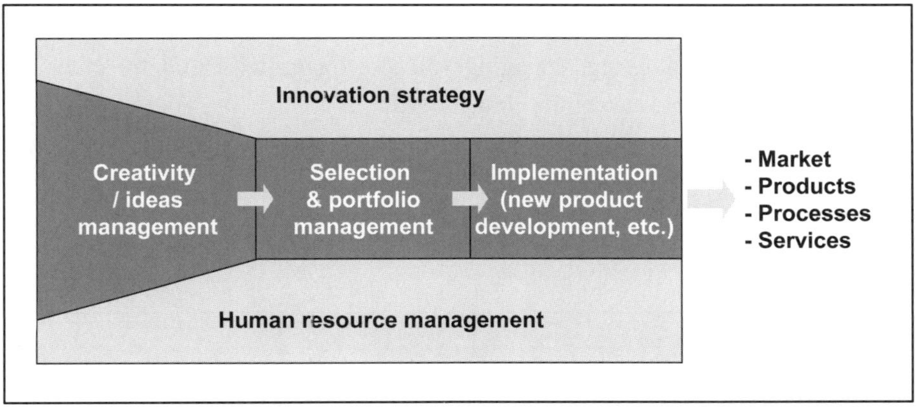

Figure 2: The Pentathlon Framework (Source: Goffin and Pfeiffer, 1999; Oke, 2007; Oke and Goffin, 2001)

The frameworks in Figures 1 and 2 outline processes for managing innovations. The challenge for service companies is how to translate these processes into their organizational processes for developing innovations. In fact research shows that unlike manufacturing companies, service companies do not typically employ a formal process for developing innovations (Bowers, 1989; Martin and Horne, 1993; Oke, 2004; Scheuing and Johnson, 1989). Many reasons have been proposed for service companies' inability to use formal processes effectively to manage their innovations. For example, it has been argued that due to their intangibility services do not readily or easily lend themselves to being managed in the same way as tangible manufactured products. However, research has shown that the use of formal processes, especially in financial services firms, is on the increase (Drew, 1995; Oke, 2002; Oke, 2007). In contrast, studies on innovation management in logistics have not attracted much attention from researchers.

Logistics innovation has been defined as "any logistics related service from the basic to the complex that is seen as new and helpful to a particular audience" (Flint et al., 2005). Flint et al. (2005) proposed an iterative process for managing logistics innovation (Figure 3). The process is comprised of four key activities:

- Setting the stage: creating and modifying environments and training people to foster innovation activities
- Customer clue gathering: engaging with customers to search for clues for changing needs and to identify unmet needs
- Negotiating, clarifying, and reflecting: a continuous improvement approach to interpreting, clarifying and negotiating customer needs
- Inter-organizational learning: engaging in joint learning and open innovation with customers

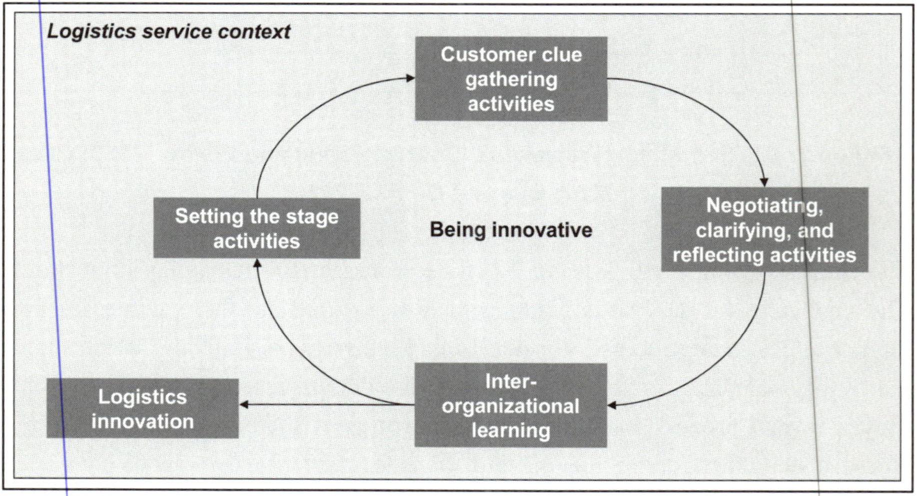

Figure 3: A Logistics Innovation Process (Source: Flint et al., 2005)

Whether logistics innovations can render themselves easily to a formal process as described in Figures 1, 2 and 3 is a different issue altogether. This is because historically, innovations by LSPs were a response to specific customer demands. While this is not peculiar to the sector, the innovations are customer-driven and are therefore more likely to be reactive rather than proactive. In order to understand the challenges that LSPs face in managing their innovations, an examination of some innovations in the sector is required.

The nature of innovations by LSPs can be associated with changes in the offerings that LSPs provide. Transportation and warehousing have traditionally been at the core of LSP offerings. Thus, it could be argued that the focus of LSPs would be on innovations that aid transportation and warehousing services. However, in the last decade or so, LSPs have had to broaden their service offerings leading to different innovative solutions.

Technology and especially information technology have played a significant role in the industry. For example, LSPs cannot afford not to have leading-edge transportation management systems and warehouse management systems to improve efficiency and customer service. Information technology innovative products such as SAP have been adopted by LSPs to provide increased visibility of movements of orders, shipments and inventories as well as co-ordination and integration of activities. More recently, LSPs have begun adopting innovations using technologies such as radio-frequency identification (RFID).

Effective tracking of movement of goods has been an important competitive criterion for LSPs. Since the introduction of a real-time tracking system for shipments by FedEx in the late 1970s, there have been significant developments and different versions of shipment tracking systems that are geared towards providing customers with better information. FedEx's new command and control system is a case in point. It comprises a relational database that uses satellite and computer communications technology to monitor routing and traffic information enabling it to coordinate FedEx logistics worldwide. Designed to act as a weather management tool, the command and control is one of FedEx's technological innovations.

UPS has also focused on technology-enabled innovations. According to Mike Eskew (UPS Chairman and CEO), "Technology gives us the capability to treat each package as if it were our only one. The 13 million packages we deliver each day are like the haystack, whereas technology allows us to treat each package like the needle."

Other than technology-based innovations, service innovations have been at the core of LSP offerings. There has been no consensus in the literature on defining service innovations. The terms 'service product innovations' and 'product innovations' have been used interchangeably to describe a particular set of innovations in service companies (Oke, 2007). On the one hand, service product innovations can be described as new developments in the core offering of service companies that tend to create new revenue streams. Service innovations, on the other hand, can be described as new developments in activities undertaken to deliver core service products (Gadrey et al., 1995; Oke, 2007). Understanding what constitutes an innovation is one of the major issues facing service organizations like LSPs.

For LSPs, technology-based innovations can be likened to product innovations in the manufacturing sector. Technology-based innovations may be tangible and are usually developed to provide improved services or service innovation. New services or service innovations that are based on technological developments are intangible in nature. For example, the FedEx call center technology enables the handling of over 500,000 telephone calls in 46 call centers across the globe daily. The capability which ensures that a high number of customer calls can be handled is the service innovation. The provision of web services by FedEx ensures customers can access information and track their packages at their own convenience. This is also a service innovation that augments the core service product of moving or shipping goods that FedEx offers. The "intelligent call routing system" which ensures that customers who need to speak to a customer representative spend a minimum amount of time waiting for a response is a tangible technology-based innovation. However, the capability provided by this technology is the service innovation. Other types of service innovations that LSPs offer to their customers may include value-added services such as product assembly and packaging, fleet management, inventory management which extend traditional logistical processes.

2.3 Innovation Barriers

In spite of the need to be innovative, several factors impede the management of innovation by LSPs. An earlier survey of senior managers in service companies (see Oke, 2004) combined with a series of informal interviews with senior logistics managers in leading LSP companies reveals the barriers to innovation management.

Lack of a Clear Definition of Innovation

Innovation means different things to different people. Interviews with senior managers within an LSP reveal that there is no agreement on what constitutes an innovation. If there is no consistent view of what an innovation is, then how can it be managed? LSPs tend to view innovation as a means to an end. In other words, an innovation in processes or technology could help to achieve a strategic objective of satisfying a customer's need in a cost-effective way. Thus, while they see innovation as key to their survival, some LSPs have yet to recognize innovation as a top strategic objective let alone devise a plan to manage it.

Reactive versus Proactive Innovations

Most of the innovations developed by LSPs tend to be reactive, in response to customer requests or needs. Reactive innovations are generally more difficult to manage than proactive innovations, probably because of the way they come about. The time frame required to deliver them is relatively shorter than that of a typical proactive innovation, hence they do not readily lend themselves to being taken through a formal management process or framework such as those described in Figures 1 and 2.

Peculiar Customers

There is a myth that LSPs are fortunate because they have a small captive audience as customers (mostly business customers) whose needs can be easily captured and translated into the development of new products and services. In reality, however, the peculiarity of LSP customers could constitute a major stumbling block to their ability to manage innovation. LSP innovations are wrapped around customer needs. Business customers want short-term gains which innovations may not deliver immediately. Business customers' main concerns are with the trade-offs among functional, service, and relationship benefits and monetary benefits (Flint et al., 2005; Lapierre, 2000; Ulaga, 2003). Thus, LSPs end up spending too much time responding to the immediate needs of the customers and not enough time working on long-term innovations (Easterby-Smith, 1997).

Ineffective Transfer of Knowledge

Just like in any innovation endeavor, knowledge transfer from customers to LSPs is important. The extent to which an LSP understands its customers' needs can determine the type of logistics innovation. On the one hand, low level learning which can be a direct response to a customer request is unlikely to produce radical innovations (Flint et al., 2005). On the other hand, high l involvement may lead to high level learning involving the exploration of multiple ways of meeting the needs of customers. Such higher levels of involvement, relationship and learning are more likely to result in new insights and innovations. The actual understanding and transferring of (sometimes tacit) knowledge is another potential barrier to innovation in LSPs. With other types of products and services that are offered to a large cross-section of customers, there are likely to be a few "hits" and "misses" depending on what different customers need. In contrast, the focused and small customer base of LSPs implies that there can only be either a "hit" or a "miss" with regards to matching customer needs. For example, an LSP representative or an account manager who deals with a specific business customer may be able to understand the customer needs but may not be able to articulate or replicate the learning process with another account manager because the knowledge may be tacit. This could impede the ability to manage innovation in the LSP.

Inability to Protect Innovations with Patents

By their very nature, some service product innovations that LSPs offer are intangible; this makes them difficult to model, describe, produce and therefore patent. For example, the core offering of movement of goods from Point A to Point B is a service product. For this reason, competitors have no problem copying this logistics service product innovation. The implication is that rather than be a first-mover in innovations, LSPs may adopt a 'wait and see' strategy. This affords them the opportunity of being able to adopt or buy in the innovation once it has proven to be a success and actually exploit it to a better effect. The inability to patent service product innovation thus serves as a barrier.

Technology as a Major Source of Innovation

Most of the innovations in the logistics industry are based on technological developments and innovations. While technology can be an enabler of innovation for LSPs, it can also be a major barrier to innovation management. Technological development or innovation is not the core competence of LSPs. Therefore, in some cases, an LSP would simply adopt an established and proven technology to improve its services. The need to manage innovation then does not arise since the innovation cannot be developed in-house. In other cases LSPs would partner with technology or information technology companies to develop a specific technology for the LSP's use.

For example, in response to regulatory pressure from the U.S. Food and Drug Administration to guarantee the temperature integrity of their drug shipments, DHL's pharmaceutical customers were clamoring for a more reliable and cost-effective solution. This would require DHL to be able to track container temperature in real time throughout the stages of transportation and delivery. Working with IBM and its partner Infratab, DHL developed an advanced temperature tracking solution that combines sophisticated sensing and RFID technology to enable the temperature-sensitive shipments to be monitored on a real-time basis.

Whether the innovation is developed jointly with (a) partner(s) or it is adopted, the innovation tends not to be owned by the LSP.

Also, there may be a heavy reliance on partner companies that help with the development or there may be a reliance on the type of technology that is available in the marketplace. According to resource dependency theory, the interdependencies among firms in a network create patterns of dependency, a situation in which firms that control valuable and scarce resources hold power over firms seeking those resources to the extent that the dependency is not mutual (Pfeffer and Salancik, 1978).

For example, greater dependence on a technology company or supplier for the development of a technology increases risk and vulnerability for the LSP since it may have fewer alternative sources of technology. The more critical the technology, the greater the dependence, and the greater the risk that the LSP faces. In spite of the potential dependency relationships risks that LSP faces, the fact that technological development is not its core competence simply limits the amount of innovation that it can engage in, let alone manage. So, while technology is an important enabler of innovation for LSPs, over-reliance on it could also turn it into a barrier and not give an LSP a sustainable competitive advantage. According to resource-based theory (Barney, 1991; Rungtusanatham et al., 2003; Wernerfelt, 1984), this is because while such technological developments are likely to be very valuable to the firm or even rare at the time of adoption, over time they are likely to become mobile, imitable and substitutable, unlike services that are developed in-house.

Lack of Effective Development Processes

Innovation involves creating new ideas and managing the process of converting those ideas to new processes, products and services. Thus, having an effective process for managing innovation is important for innovation performance (De Brentani, 1989; Mitchell Madison Group, 1995; Mohammed-Salleh and Easingwood, 1993). The development process for service innovations tends to be complex because the intangible nature of services makes it difficult for service organizations such as LSPs to develop new products and services systematically

(Edgett, 1996; Edvardsson, et al., 1995; Griffin, 1997; Martin and Horne, 1993). Where the focus is on technology innovation that depends on external partners, the development process would depend on integrating LSP processes with the partnering companies' processes – which might turn out to be a difficult task. A lack of processes could imply that innovations are done in an ad-hoc and an informal manner so it is hard to manage subsequent innovations and replicate successful innovations (Flint et al., 2005; Ljungberg and Larsson, 2001; Oke, 2004).

Difficulty in Concept Testing

Due to the intangibility of services, concept testing has been problematic for service organizations (Bowers, 1989; Easingwood, 1986; Edgett, 1993). However, some service organizations have been able to overcome this barrier. For example, when offering new services, the Bank of America is able to select a few test branches. Test services are offered to customers in parallel to the normal range of services by staff who have been trained to offer these new services and monitor customers' reactions (Thomke, 2003). LSP's business-to-business relationships typically would have few but focused customers. This may mean that the opportunity to test new services with these few customers may be more limited and more risky than in a business-to-consumer relationship where there is a large consumer base unless there is a very close relationship between the LSP and its business customers.

2.4 Implications and Recommendations

The discussion in this chapter should help LSPs appreciate the challenges that they may face in managing innovation. Although LSPs offer a unique type of services, the challenges faced by LSPs in managing innovation are not dissimilar to the challenges faced by other service companies such as financial services companies. With the growth of the logistics sector and with increasing competitive pressure, the need for LSPs to provide innovative services cannot be overemphasized. A number of potential barriers to innovation management are identified and discussed. Because LSPs tend to operate under a strictly customer-centric approach of delivering short-term solutions for short-term gains, they may be in danger of ignoring innovations that could help them to compete more effectively. This study offers insights into factors that LSPs should look out for if they are to become more innovative. The study makes four recommendations for LSPs in terms of overcoming the barriers to innovation management:

- Managing innovation requires a clear definition of innovation. There must be a common and a consistent understanding of the meaning of innovation across the organization. Innovation should not only be seen as a means to an end but also as an end in itself, capable of delivering differentiated services to customers. Hence innovation ought to form an integral part of business strategy. A definition of innovation will emphasize the relative importance of the focus areas for innovation for LSPs that may include technological developments, service and service product innovations. This may mitigate the barrier caused by an over-reliance technological innovation. LSPs would be unable to ignore developments in other innovations once the focus areas have been established through a definition of what innovation means to LSPs.
- It is vital to develop a process for managing innovations in LSPs. There should be different processes for different types of innovation, hence a flexible process structure is recommended. For example, reactive innovations may require a shorter and less formal process than proactive innovations. Processes would help to overcome the ad-hoc and informal development issues that often complicate the managing of subsequent innovations and replicating successful ones.
- Having a process ensures accountability and consistency in terms of what is offered to customers. This is especially important given the intangible nature of services. Performance measures can then be based on the processes defined. Installing performance measures is vital for effective innovation management as it identifies areas where opportunities for improvements may exist.
- LSPs tend to depend on technological developments for their innovations. However, developing technology is not the core competence of LSPs, hence they may either buy-in the technology or partner with technology companies. Whichever option is taken, the innovations tend to be easily and readily available to competitors. In fact, competitors that adopt a "wait and see"

strategy may profit more from such innovations as they are able to improve on them. As such LSPs need to increase their focus on service innovations which although cannot be patented, but they cannot easily be copied since they are less obvious to competitors than technology-related innovations.

As more operations become globalized and outsourced, the demand for the services provided by logistics companies will continue to increase. Innovation will remain a key competitive criterion in the sector. In the same way, the challenges to manage innovation will persist. Only those LSPs that are able to identify and overcome the challenges in order to manage innovation in a systematic way will become the leaders in the sector.

References

Barney, J.B. (1991). Firm Resources and Sustained Competitive Advantage, *Journal of Management,* 17 (1), 99-120.

Booz, Allen and Hamilton (1982). *New Products Management for the 1980s*, New York: Booz, Allen and Hamilton.

Bowers, M.R. (1989). Developing New Services: Improving the Process Makes it Better, *Journal of Services Marketing*, 3 (1), 15-20.

De Brentani, U. (1989). Success and Failure in New Industrial Services, *Journal of Product Innovation Management*, 6 (4), 239-258.

Delaney, R.V. (1999). *10th Annual State Logistics Report.* Available: http://www.cassinfo.com.

Drew, S. (1995). Strategic Benchmarking: Innovation Practices in Financial Institutions, *International Journal of Bank Marketing*, 13 (1), 4-16.

Easingwood, C.J. (1986). New Product Development for Service Companies, *Journal of Product Innovation Management*, 3 (4), 264-275.

Easterby-Smith, M. (1997). Disciplines of Organizational Learning: Contributions and Critiques, *Human Relations*, 50 (9), 1085-1113.

Economist Intelligence Unit (2006). EIU Country – United Kingdom, April, 2006, London.

Edgett, S. (1993). Developing New Financial Services within UK Building Societies, *International Journal of Bank Marketing*, 11 (3), 35-43.

Edgett, S. (1996). The New Product Development Process for Commercial Financial Services, *Industrial Marketing Management*, 25 (6), 507-515.

Edvardsson, B.; Haglund, L.; Mattsson, J. (1995). Analysis, Planning, Improvisation and Control in the Development of New Services, *International Journal of Service Industry Management*, 6 (2), 24-35.

Flint, D.J.; Larsson, E.; Gammelgaard, B.; Mentzer, J.T. (2005). Logistics Innovation: A Customer Value-Oriented Social Process, *Journal of Business Logistics*, 26 (1), 113-147.

Gadrey, J.; Gallouj, F.; Weinstein, O. (1995). New Modes of Innovation: How Services Benefit Industry, *International Journal of Service Industry Management*, 6 (3), 4-16.

Goffin, K.; Pfeiffer, R. (1999). *Innovation Management in UK and German Manufacturing Companies*, London: Anglo-German Foundation.

Griffin, A. (1997). PDMA Research on New Product Development Practices: Updating Trends and Benchmarking Best Practices, *Journal of Product Innovation Management*, 14 (6), 429-458.

Lapierre, J. (2000). Customer-Perceived Value in Industrial Contexts, *Journal of Business and Industrial Marketing*, 15 (2), 122-140.

Ljungberg, A.; Larsson, E. (2001). *Processbaserad Verksamhetsutveckling*, Lund, Sweden: Studentlitteratur.

Martin, C.R.; Horne, D.A. (1993). Services Innovation: Successful Versus Unsuccessful Firms, *International Journal of Service Industry Management*, 4 (1), 48-64.

Mitchell Madison Group (1995). *New Product Development: Organization and Process in Service Companies*, New York: Mitchell Madison Group.

Mohammed-Salleh, A.; Easingwood, C. (1993). Why European Financial Institutions Do Not Test-Market New Consumer Products, *International Journal of Bank Marketing*, 11 (3), 23-28.

Oke, A. (2002). Improving the Innovative Capability of a Service Company, *Journal of Change Management*, 2 (3), 272-281.

Oke, A. (2004). Barriers to Innovation in Service Companies, *Journal of Change Management*, 4 (1), 31-44.

Oke, A. (2007). Innovation Types and Innovation Management Practices in Service Companies, *International Journal of Operations and Production management*, 27 (6), 564-587.

Oke, A.; Goffin, K. (2001). Innovation Management in the Service Sector. Management Focus, *Journal of the Cranfield School of Management*, UK, Summer Issue.

Pfeffer, J., Salancik, G.R., (1978). *The External Control of Organizations*, New York: Harper & Row.

Rungtusanatham, M., Salvador, F., Forza, C., Choi, T. (2003). Supply-chain Linkages and Operational Performance: A Resource-Based Perspective, *International Journal of Operations and Production Management*, 23, 1084-1099.

Scheuing, E.E.; Johnson, E.M. (1989). New Product Development and Management in Financial Institutions, *International Journal of Bank Marketing*, 7 (2), 17-21.

Thomke, S. (2003). R&D Comes to Service, *Harvard Business Review*, 81 (4), 71-79.

Ulaga, W. (2003). Capturing Value Creation in Business Relationships: a Customer Perspective, *Industrial Marketing Management*, 32 (8), 677-693.

Wernerfelt, B. (1984). A Resource-Based View of the Firm, *Strategic Management Journal*, 5 (2), 171-180.

Chapter 3

Innovative LSP Services: The Marketing Challenge

Thomas W. Speh

Abstract

The growing market for logistics business services poses special challenges and opportunities for the marketing manager of a logistics service provider (LSP). This chapter will explain the unique aspects of marketing LPS services and identify the unique challenges faced by companies as they create marketing strategies for those services.

Unlike tangible products, business services have unique dimensions which affect the way in which they are marketed. When the services are new, the challenges become even more daunting. This chapter will provide an understanding of how the unique aspects of new services affect the firm's marketing approach. The chapter will highlight the role of service quality, customer satisfaction, and loyalty in the success of service marketing. In addition, the chapter will explain the factors that must be considered in formulating a service marketing strategy for new and innovative offerings provided by LSPs. The chapter will conclude by examining the determinants of new service success and failure.

3 Innovative LSP Services: The Marketing Challenge

3.1 Introduction

Business services have become key elements for almost every economy in the world. The world trade in services, which is growing faster than the world trade in goods, has created a huge global market for service workers, as over 1.4 billion people are employed in service jobs worldwide (finfacts.com, 2005). Service workers represent approximately 51% of the global workforce (ILO, 2005). The dramatic growth in the service sector is reflected in the growth of the logistics service provider (LSP) industry. The LSP industry is developing rapidly as a result of the emerging demand for advanced logistics services (Hertz and Alfredsson, 2003). This surge in demand for LSP services can be linked to a variety of factors, including globalization, desire to reduce order cycle time, cost pressures, increased customer orientation, desire to focus on core competencies, the need to integrate supply chains, and more effective asset utilization.

As the demand for the services offered by the LSP industry has increased, so has the need to create effective marketing strategies to capitalize on the demand. Although firms are outsourcing key logistics processes at a record pace, LSPs cannot assume that business will simply come to them. The most successful LSPs have crafted and executed effective marketing strategies, including the design, promotion and distribution of new services, and properly priced their services. Adding to the challenge of designing a strong marketing strategy is the fact that many LSPs are constantly offering unique services. The fact that the firm is marketing a service, and often an innovative one, makes the creation of effective marketing strategies even more difficult.

The marketing of business services is complex for a variety of reasons, not the least of which is the fact that a business service is often nebulous and intangible. When the service is quite different from existing ones, the LSP will be called upon to formulate a strategy that recognizes and deals with the innovative nature of that service.

This chapter will explore why LSP services are difficult to market, particularly when they are new and different. It will then examine the process for creating a marketing strategy for innovative LSP services, focusing on an integrated mix of marketing tools for creating a distinctive strategy to potential customers. Finally, the chapter will explore the inherent problems in formulating marketing strategy for "new" or innovative services, and how these problems can be solved.

In recent years, marketing of services has grown significantly. Many business service providers, including LSPs, have been reluctant to actively and aggressively market their services, often because they are too focused on provision of the service, than on how best to reach out to potential customers through an integrated marketing strategy. Quite simply, many LSPs do not understand the role of marketing in acquiring new clients. However, with the rise of competition in the LSP services industry and with the significant increase in the outsourcing of key activities, LSPs must develop highly focused marketing programs. Marketing strategies for LSP services will be based on understanding the key aspects of a business service and how those key aspects drive the creation and execution of marketing strategy. Further, if the service is new and innovative, formulation of the marketing strategy takes on another set of key dimensions.

A further challenge to the LSP is that many LSPs have little understanding of and only a slight focus on marketing. Because logistics is complicated and involves a deep understanding of many different processes, the typical LSP often places significant emphasis on execution of these processes. The emphasis and training in marketing is generally weak. Hence, the LSP must first understand the cornerstones of marketing strategy, how to develop the elements of their marketing strategy, and finally to execute the strategy. Much of the chapter will focus on the critical aspects of marketing strategy: understanding the customer, creating services that provide benefits to those customers, and then executing a marketing strategy composed of the service "product," the selling, promoting, pricing and "delivering" the service product.

There are inherent differences between goods and services. Put simply, services are deeds, processes, and performances. For example, the core offerings of a management consultant are primarily deeds and actions performed for customers. LSPs provide a huge array of possible services, many of which are processes linked to the timely and cost-effective delivery of products to the client's customers. The most basic and universally recognized difference between goods and services is *intangibility*. Services are more intangible than manufactured goods, and manufactured goods are more tangible than services. Because services are actions or performances, they cannot be seen or touched as tangible goods can be.

3.2 LSP Services: Distinguishing Characteristics

Figure 1 provides a useful tool for understanding the product-service definitional problem. The continuum suggests that there are very few *pure products* or *pure services*. For example, a laptop computer is a physical object made up of tangible elements that facilitate the work of an individual and an organization. In addition to the physical design and performance characteristics of the computer, the quality of technical service support is an important dimension of the marketing program. Thus, most market offerings comprise a combination of tangible and intangible elements.

The services offered by LSPs mix tangible and intangible elements. A third-party warehousing firm has many tangible assets it uses in providing service to its clients – buildings, racks, fork lifts, pallets, labels, scanners and the like. However, the output produced by all those assets – the timely delivery of undamaged product to the right customer with the proper assortment and quantities of product – is not so tangible. As the LSP attempts to market these services to a new customer, it must find ways to make the intangible elements – on time delivery, for example, more tangible. In this case, performance records of on-time delivery to all the firm's customers would show prospective customers that the delivery performance has been achieved, thus making the intangible service time significantly more real. In similar fashion, testimonials from current customers regarding on-time delivery serve the same purpose. The chapter will explore this issue in a later section on the creation of marketing strategies.

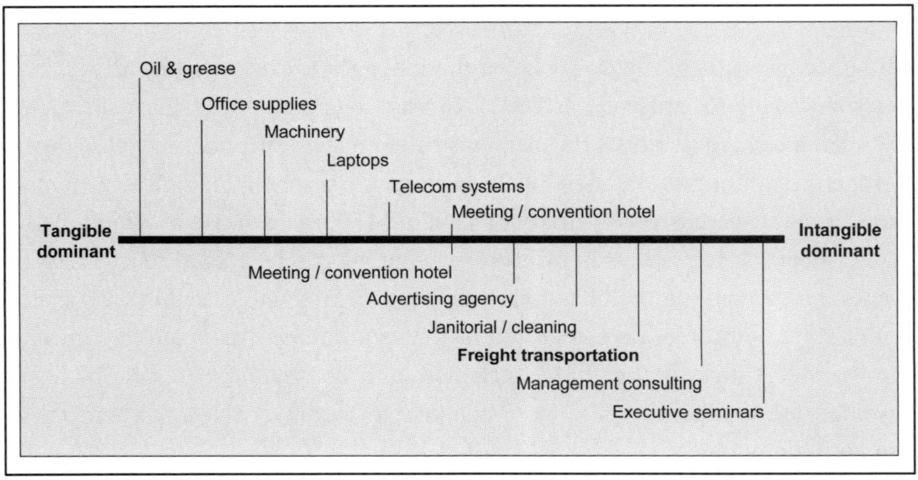

Figure 1: Business Services Classification Based on Tangibility

Whether an offering is classified as a good or as a service depends on how the organizational buyer views what is being bought – whether the tangible or the intangible elements dominate. On one end of the spectrum, grease and oil are tangible-dominant, and the essence of what is being bought is the physical product. A management seminar dealing with inventory management, in contrast, is intangible-dominant because what is being bought – professional development, education, learning – has few, if any, tangible properties. A convention hotel is in the middle of the continuum because the patron will receive an array of both tangible elements (meals, beverages, notepads) and intangible benefits (courteous personnel, fast check-ins, meeting room ambiance). The hotel is very similar to an LSP, as the company provides both tangible and intangible benefits.

The concept of tangibility is especially useful to the marketer of LSP services because their offerings combine product and service elements. The key management task is to evaluate (from the buyer's standpoint) which elements of the offering dominate. The more the market offering is characterized by intangible elements, the more difficult it is to apply the standard marketing tools that were developed for products. The LSP marketer must focus on specialized marketing approaches appropriate for services.

The concept of tangibility also helps the manager to focus on the firm's *total market offering* (Sharma et al., 2001). In addition, it helps the manager recognize that a change in one of the elements of the market offering may completely change the offering in the view of the customer. For example, an LSP who proposes to hold spare parts inventory at a central location and use overnight delivery to meet the requirements of a prospective client's customer, must focus its marketing strategy on the intangible nature of the intangible benefits associated with reduced customer inventory and fast transportation. This relatively intangible "service," must be carefully explained and the intangible results of lower inventory costs must be made more concrete to the buyer through a persuasive promotion program.

In summary, an LSP's services are those market offerings that are intangible-dominant. However, few services are totally intangible. An LSP will need to place primary marketing focus on the intangible elements of its market offerings because these elements are most difficult for prospective customers to "feel" and understand.

In addition to the tangibility criterion, business services have other important distinguishing characteristics that influence their marketing. Figure 2 summarizes the core characteristics of business services.

Because most services are *consumed as they are produced*, a critical element in the buyer-seller relationship is the effectiveness of the individual (the warehouse order picker, the package delivery service driver, the network design consultant) who actually provides the service. From the service firm's perspective, the entire marketing strategy may rest on how efficiently the individual service provider interacts with the customer. Here the actual service delivery takes place and the promise to the customer is either kept or broken. This critical point of contact with the customer is referred to as interactive or real-time marketing (Sharma et al., 2001). As a result, the recruiting, hiring, and training of personnel assume special importance in LSPs. It is likely that many LSPs do not think of their hiring decisions as a key element of their marketing offering, but because LSP personnel often make final contact with the customer, hiring decisions are indeed critical aspects of the LSP's marketing approach.

Characteristics	Examples	Marketing implications
Simultaneous production and consumption	Telephone conference calls; management seminar, repair of equipment	Direct-seller interaction requires that service be done "right"; requires high-level training for personnel' requires effective screening and recruitment
Non-standardized output	Inventory management systems varies with the individual LSP; product damage varies from shipment to shipment	Emphasizes strict quality control standards; develop systems that minimize deviation and human error; prepackage the service; look for ways automate
Perishability; inability to store or stockpile	Unfilled space in a container or truck; an idle order picker; unrented warehouse space	Plan capacity around peak demand; use pricing and promotion to even out demand peaks and valleys; use overlapping shifts for personnel
Lack of ownership	Use of a railroad car; use of a consultant's know-how; use of a mailing list	Focus promotion on the advantages of non-ownership; reduced labor, overhead, and capital; emphasize flexibility

Figure 2: Unique Service Characteristics

Figure 2 shows that the service offering is *non-standardized,* meaning that the quality of the service output may vary each time it is provided. Services vary in the amount of equipment and labor that is used to provide the service. For example, a significant human element is involved in teaching an executive seminar but not in providing overnight airfreight service. Generally, the more labor involved in a service, the less uniform the output. In these labor-intensive cases, the user may also find it difficult to judge the quality *before* the service is provided. Because of uniformity problems, business service providers must focus on finely-tuned quality-control programs (e.g., six sigma programs), invest in "systems" to minimize human error, and find ways of automating the service. Each of these approaches is designed to reduce the variability in performance. LSPs are particularly susceptible to variability of service performance because of the complexity of the job and the inherent uncertainties associated with consumer demand. For these reasons, special attention must be paid to standardizing service performance, and this uniform level of performance needs to be measured and tracked, so that it can be corrected and adjusted to customer requirements.

A significant issue in regards to standardizing service performance arises when the LSP has created a new, innovative service offering. Here, there is no historical performance to use as evidence of uniformity in service outcomes, so the firm must go to great lengths convince the prospective client that the services will be consistent and then assure this consistency. Often it will require that the LSP "gold plate" the new service offering, that is, invest substantially more in the new service than might be necessary simply to assure that the initial offering of the new service will not fail. For example, when offering a new cross-dock service to a client, it may be necessary to add labor to the operation, double- and triple-check orders and to provide backup inventory in case inbound and outbound schedules do not match.

Generally, services *cannot be stored*; that is, if they are not provided at the time they are available, the lost revenue cannot be recaptured. Tied to this characteristic is the fact that demand for services is often unpredictable. The LSP service provider must carefully evaluate capacity – in a service business, capacity is a substitute for inventory. If capacity is set for peak demand, a "service inventory" must exist to supply the highest level of demand. As an example, many LSPs manage just-in-time delivery of parts to a client's production line. If, on any given day, the demand for components exceeds normal expectations, the provider would be in jeopardy of causing a production shut down. So, the answer is to keep some buffer inventory for when demand exceeds the normal expectation. Although JIT systems are predicated on eliminating inventories throughout the supply chain, the LSP provider may opt to set some level of inventory in excess of the normal expectation, so that no manufacturing plant will ever be shut down due to a lack of available components. Obviously, setting high capacity levels is costly, and the firm must analyze the cost versus the lost revenue and customer goodwill that might result from maintaining lower capacity.

The final significant dimension of services shown in Figure 2 is the fact that the service buyer uses, but *does not own*, the service purchased. Essentially, payment for a service is a payment for the use of, access to, or hire of items. As a result, the LSP must feature the advantages of non-ownership in its communications to the marketplace. The key benefits to emphasize are reductions in staff, overhead, and capital associated with a third party providing the service.

Although there may be exceptions, these characteristics of business services are a useful framework for understanding the underlying nature of the services and for isolating special marketing strategy requirements. The framework suggests that LSPs should pursue different types of strategies – due to the intangibility and heterogeneity of their services. In this case, providers of key services should develop marketing strategies that emphasize word-of-mouth communication, provide tangible evidence, and employ value pricing to overcome the problems posed by intangibility and heterogeneity (Clemes et al., 2000).

3.3 Service Quality

Quality standards are ultimately defined by the LSP's customers. Actual performance by the service provider or the provider's perception of quality is of little relevance compared to the customer's perception of the service. "Good" service results when the service provider meets or exceeds the customer's expectations. As a result, many management experts argue that LSPs should carefully position themselves so that customers expect a little less than the firm can actually deliver. The strategy: "under promise" and "over deliver."

Because many of the services provided by LSPs are intangible and non-standardized, buyers tend to have greater difficulty in evaluating services than in evaluating goods. The inability to depend on consistent service performance and quality may lead buyers to experience more perceived risk. As a result, buyers utilize a variety of pre-purchase information sources to reduce risk. Information from current users (word of mouth), the use of consultants to evaluate service providers, significant and lengthy due diligence are particularly important. In addition, the evaluation process for services tends to be more abstract, more random, and more heavily based on symbology rather than on concrete decision variables. LSPs must consider all of these elements as they formulate their marketing strategy, particularly when they are dealing with new, highly innovative services. Importantly, LSP's may need to spend more time pleasing and responding to consultants than they do responding to the client, at least in the initial phases of securing new business. They must realize that the decision process may be lengthy and be prepared to provide extensive data, often very sensitive cost data, in order to meet the prospective buyer's due diligence requirements.

Research provides some valuable insights into how customers evaluate service quality. According to Figure 3, customers focus on five dimensions in evaluating service quality: reliability, responsiveness, assurance, empathy, and tangibles.

Dimension	Description	Examples
Reliability	**Delivering on promises**	**Promised delivery time met**
Responsiveness	**Being will to help**	**Prompt reply to customers' requests**
Assurance	**Inspiring trust and confidence**	**Professional and knowledgeable staff**
Empathy	**Treating customer as Individuals**	**Adapt to unique needs of the customer**
Tangibles	**Representing the service physically**	**Distinctive materials; brochures documents**

Figure 3: The Dimensions of Service Quality (Source: Zeithaml et al., 2006)

Among these dimensions, reliability – delivery on promises – is the most important to customers. High-quality service performance is also shaped by the way in which the service is provided by frontline service personnel. To the customer, service quality is equated with a responsive employee, one who inspires confidence, one who adapts to the individual needs or preferences of the customer and delivers the service in a professional manner. In fact, the performance of employees who are in contact with the customer may compensate for temporary service quality problems. For example, a problem occurs with packaging quality, resulting in higher damage rates than expected. By promptly acknowledging the error and responding quickly to the problem, the LSP representative may even strengthen the firm's relationship with the customer.

Customer satisfaction and loyalty

Extensive research has shown that four components of a firm's offering and its customer-linking processes affect customer satisfaction:

1. The basic elements of the service that customers expect all competitors to provide
2. The basic support services, such as technical assistance or training, that make the service more effective or easier to use
3. A recovery process for quickly fixing product or service problems
4. Extraordinary services that so excel in solving customers' unique problems or in meeting their needs that they make the product or service seem customized (Jones and Sasser, 1995).

Leading service firms carefully measure and monitor customer satisfaction because it is linked to customer loyalty and, in turn, to long-term profitability. Xerox, for example, regularly surveys more than 400,000 customers regarding service satisfaction using a five-point scale. In analyzing the data, Xerox executives made a remarkable discovery: *very* satisfied customers (those who gave a five rating) were far more loyal than satisfied customers. Very satisfied customers, in fact, were *six times* more likely to repurchase Xerox products than were satisfied customers (Heskett et al., 1994).

Service Recovery

It is not possible for an LSP to provide flawless service in every transaction. However, the way in which the firm responds to a client's service problems has a crucial bearing on customer retention and loyalty. Service recovery encompasses the procedures, policies, and processes that a firm uses to resolve customer service problems promptly and efficiently. For example, when IBM receives a customer complaint, a specialist, who is an expert in the relevant service area, is assigned as "resolution owner" of that complaint. Upon being assigned a customer complaint or problem, the IBM specialist must contact the customer within 48 hours (except in the case of severe problems where the required response is made much faster). Larry Schiff, a marketing strategist at IBM, describes how the process proceeds from there:

"*They introduce themselves as owners of the customer's problem and ask: What's it going to take for you to be very satisfied with the resolution of this complaint? . . . Together with the customer, we negotiate an action plan and then execute that plan until the customer problem is resolved. The problem only gets closed when the customer says it is closed, and we measure this as well*" (Schiff, 2001).

Service providers who satisfactorily resolve service failures often see that their customer's perception of service quality rises. One study in the ocean freight shipping industry found that clients who expressed higher satisfaction with claims handling, complaint handling, and problem resolution have a higher overall level of satisfaction with the shipping line (Durvasula et al., 2000). Therefore, LSPs should develop thoughtful and highly responsive processes for dealing with service failures. Service failure strategies must be created and be ready to be acted upon, as some studies have shown that customers who experienced a service failure and had it corrected to their satisfaction have greater loyalty to the supplier than those customers who did not experience a service failure.

The quality of service provided to business customers has a major effect on customer "defections" – customers who do no extend a contract at the end of its term. Service strategists point out that customer defections have a powerful impact on the bottom line (Reichheld, 2001). As a company's relationship with a customer lengthens, profits rise – generally considerably. For example, one service firm found that profit from a fourth-year customer is triple that from a first-year customer. Many additional benefits accrue to service companies that retain their customers: They can charge more, the cost of doing business is reduced, and the long-standing customer provides "free" advertising. The implications are clear: service providers should carefully track customer defections and recognize that continuous improvement in service quality is not a cost, but, says Frederick Reichheld "an investment in a customer who generates more profit than the margin on a one-time sale".

Return on Quality

A difficult decision for the business services marketing manager is to determine how much to spend on improving service quality. Clearly, expenditures on quality have diminishing returns – at some point, additional expenditures on service quality do not increase profitability. To make wise decisions on the level of expenditures on quality, managers must justify quality efforts on a financial basis, knowing where to spend on quality improvement, how much to spend, and when to reduce or stop the expenditures. Researchers have developed a technique for calculating the "return on investing in quality" (Rust et al., 1995). Under this approach, service quality benefits are successively linked to customer satisfaction, customer retention, market share, and, finally, to profitability. The relationship between expenditure level and customer satisfaction change is first measured by managerial judgment and then through market testing. When the relationship has been estimated, the return on quality can be measured statistically. The significant conclusion is that quality improvements should be treated as investments. They must pay off, and spending should not be wasted on efforts that do not produce a return.

3.4 Marketing Strategy for LSP Firms

To meet the needs of service buyers, an integrated marketing strategy is required. First, target segments must be selected, and then a marketing mix must be tailored to the expectations of each segment. The key elements of the service marketing mix include the development of the service "product" or "service package", pricing, promotion, and distribution. Each requires special consideration by the LSP marketing manager.

In terms of the approach that firms develop to interact with their customers, a recent study has shown that business-to-business service firms are more likely to emphasize *relationship* strategies as opposed to *transactional* strategies. Since the transactional mode involves an arms-length relationship, success in marketing business services hinges on the business marketer's ability to develop close and long-lasting ties with customers – based on buyer-seller dependence. Importantly, the emphasis for the LSP is on managing the total buyer-seller interaction process.

As with any marketing situation, development of the marketing mix will be contingent upon the customer segment to be served. Every facet of the service to be offered, as well as the methods for promoting, pricing, and delivering the service, will hinge upon the needs of a reasonably homogeneous group of customers. The process for segmenting the market for LSP services is unique. First, market segments are often very narrow. This situation reflects the fact that many customers expect services to be customized. Expectations may not be met if the service received is standardized and routine. In many cases, particularly for new, innovative LSP services, the market segment simply may be one customer with a very unique need. Second, LSP service segmentation focuses on what the buyers expect, not on what they need. The assessment of buyer expectations will play a very large role in selecting a target market and developing the appropriate service package (Figure 4). This assessment is critical because so many studies have shown large differences between the ways that customers define and rank different service activities and the ways suppliers do.

Because expectations play such an important role in determining ultimate satisfaction with a service, service quality expectations can be used to segment markets for LSP services. One study in the mainframe software industry revealed significant differences between "software specialists" (software experts) and "applications developers" (users of software) in the same firm regarding their expectations of new software. (Pitt et al., 1996) The developers (users) had higher expectations regarding the quality of the supplier's equipment, the responsiveness of their employees, and the amount of personal attention provided. The study concluded that different buying center members may well have different perspectives and expectations of service quality. On the basis of this example from the software industry, the LSP should evaluate service quality expectations when designing a marketing strategy.

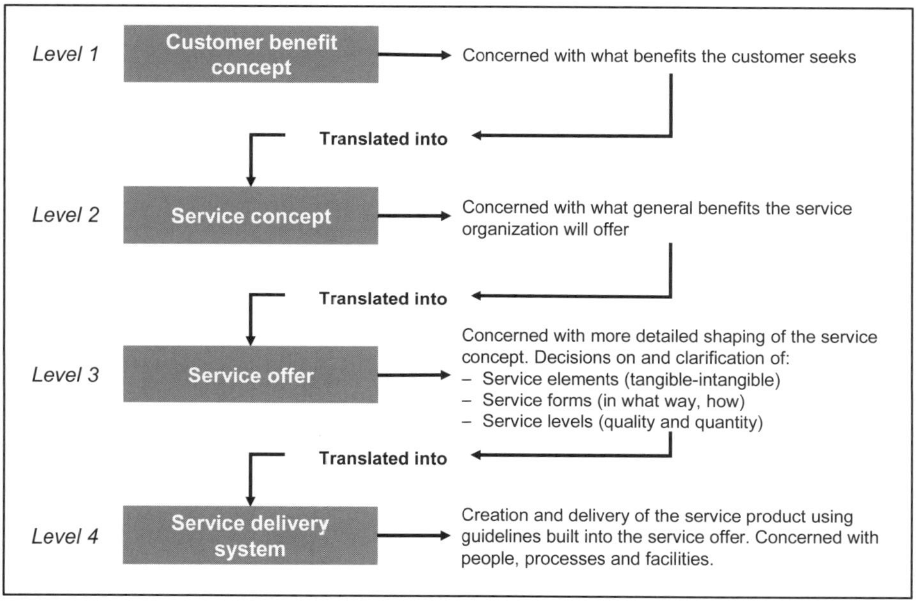

Figure 4: Conceptualizing the Service Product/Package (Source: Cowell, 1984)

Finally, segmenting service markets helps the firm to adjust service capacity. Segmentation will usually reveal that total demand is made up of numerous smaller, yet more predictable, demand patterns. For a warehouse, the demand patterns of a cross-dock customer, a pure storage customer, a handling-intensive customer can all be forecast individually and capacities adjusted for each segment's demand pattern.

The Service "Package"

The service package can be thought of as the "product" dimension of service, including decisions involving the essential concept of the service, the range of services provided, and the quality and level of service. In addition, the service package must consider some factors that are unique to services – the personnel who perform the service, the physical product that accompanies the service, and the process of providing the service. A useful way of conceptualizing the service product is shown in Figure 4.

Customer-Benefit Concept

Services are purchased because of the benefits they offer, and a first step in either creating a new service or evaluating an existing one is to define the customer-benefit concept – that is, evaluate the core benefit that the customer will derive from the service. This process is absolutely critical in the marketing of new, innovative LSP services. An understanding of the customer-benefit concept will focus the LSP's attention on those attributes – functional, effectual, and psychological – that must be not only offered but also tightly monitored from a quality-control standpoint. For example, a manager selecting a trucking company for deliveries to the stores of his/her client is purchasing a core benefit that could be stated as "on-time availability." The LSP trucker must then assess the full range of service attributes and components necessary to provide a successful "delivery." Obviously, a wide variety of service elements will come into play: (1) delivery to the correct store; (2) damage-free product; (3) courtesy of the driver; (4) proper documentation and advanced shipping notification; and (5) staff responsiveness if a problem arises. These are all factors that will be important to both the client and particularly to the client's customer. As another example, Federal Express Custom Critical does not provide its customers with "transportation," per se. Its customer-benefit concept focuses on absolute conformance to the time-sensitivity of the delivery.

Service Concept

Once the customer-benefit concept is understood, the next step is to articulate the service concept, which defines the general benefits the LSP will provide in terms of the bundle of services sold to the customer. The service concept translates the customer-benefit concept into the range of benefits the LSP will *provide*. For a package delivery firm, the service concept might specify the benefits that it will develop: flexibility, responsiveness, and courtesy of drivers; a full range of tracking capabilities; flexible schedules; pre-arranged return shipping if required; professional personnel; and climate-controlled vehicles (where necessary).

Service Offer

Intimately linked with the service concept is the service offer, which spells out those services to be offered; when, where, and to whom they will be provided; and how they will be presented. The service elements that make up the total service package, including both tangibles and intangibles, must be determined. The service offer of the contract warehouse includes a multitude of tangible elements (handling equipment, surveillance cameras, storage rack, pest control devices, scanners, computing equipment, and pallets.) and intangible elements (attitude of the warehouse general manager, warmth of greetings from clerks and administrative staff, response to unique requests, IT sophistication). Generally, management will find it easier to manage the tangible (equipment and physical) elements of the service than to control the intangible elements.

Service Delivery System

The final dimension of the service product is the service delivery system – how the service is provided to the customer. The delivery system includes appropriately conceived jobs for people; personnel with capabilities and attitudes necessary for successful performance; equipment, facilities, and layouts for effective customer work flow; and carefully developed procedures and processes aimed at a common set of objectives. Thus, the service delivery system should provide a well-designed blueprint that describes how the service is rendered for the customer.

For physical products, manufacturing and marketing are generally separate and distinct activities; for services, the two are often inseparable. The service performance and the delivery system both create the product and deliver it to the customers. This feature of services underscores the important role of people, particularly service providers, in the marketing process. Drivers, warehouse GM's, customer service representatives, IT technicians, supervisors and repair people are often intimately involved in customer contact, and they decidedly influence the customer's perception of service quality. As a result, LSPs, have to pay close attention to both people and physical evidence (tangible elements such as uniforms, cleanliness of buildings and vehicles, and signage) when designing the service package to present to the customer.

Service Personnel

A crucial step in creating an appealing service package is to ensure that the customer-benefit concept is known, understood, and accepted by all personnel. As one service marketing expert states, "[s]o important are people and their quality to organizations and […] services that 'internal marketing' is considered to be an important management role to ensure that all staff are customer conscious." In short, the attitudes, skills, knowledge, and behavior of service personnel have a critical impact on the customer's satisfaction with the service.

Pricing LSP Services

Each type of LSP business will have its own technical considerations regarding the way in which a price for a particular service is derived. For example, the pricing of a complex warehousing operation involving thousands of orders per month and wide array of SKUs will take many days if not weeks to derive, and input will be required from a vast network of managers, accounting personnel, and technicians. In contrast, pricing of transportation of a single shipment from Point A to Point B may simply be derived from a straightforward price sheet. Hence, the focus here will be to derive important generalizations concerning the pricing of LSP services and the unique characteristics of services that are responsible for creating some special pricing problems and opportunities.

Perishability and Managing Demand/Capacity

The demand for services is rarely steady or predictable enough to avoid service perishability. An extremely difficult decision for a marketer of LSP services is to determine the capacity (inventory) of the system: should it meet peak demand, average demand, or somewhere in between? Pricing can be used to manage the timing of demand and align it with capacity levels, although this is difficult when contracts are signed for long periods based on total movement of products over that time frame. This is particularly true of warehousing services and transportation services where the demand for space in the warehouse and for shipments can be erratic and unpredictable from day to day. Hence, it may be necessary to consider a pricing scheme where surcharges are added when demand exceeds some extreme level.

To manage demand, the LSPs may offer off-peak pricing schemes and price incentives for service orders that are placed in advance. For example, a transport firm might develop price incentives for shipping larger quantities at off-peak seasons or shipments that are picked up at a warehouse between 10 pm and 8 am. Alternatively, lower rates may be offered if a steady stream of loads is guaranteed to be offered in a specified period of time. Similarly, it may also be possible, depending on demand elasticity and competition, to charge premium rates for services provided at peak demand periods. It is interesting to note, however, that one study of strategies utilized by service firms showed that many service firms do not reduce prices to increase business during slow periods (Zeithaml et al., 1985).

Service Bundling

Many business services include a core service as well as a variety of peripheral services. How should the services be priced – as an entity, as a service bundle, or individually? Bundling is the practice of marketing two or more services in a package at a special price. Bundling makes sense in the business service environment because most service businesses have a high ratio of fixed to variable costs and a high degree of cost sharing among their related services. For example an LSP warehousing firm might invest in packaging equipment, stretch wrap machines, labeling equipment and of course, the normal number of lift trucks and other handling devices. The marginal cost of providing additional services (specialized packaging, labeling, assembly) to the core service (basic warehouse space and order picking) is generally low. Hence, offering the typical warehouse space and handling along with the specialized packaging and labeling at a very attractive single rate may provide a strong incentive to the client in addition to offering a reasonable profit to the LSP.

A key decision for the service provider is whether to provide pure or mixed bundling. In pure bundling, the services cannot be purchased separately. In mixed bundling, the customer can either purchase one or more services individually or purchase the bundle. For example, an LSP warehouse firm can provide its services – storage, product handling, and clerical activities – in a price-bundled form by charging a single rate (five cents) for each case received by the warehouse from its manufacturer-client. Or the firm may market each service separately and provide a rate for each service individually (two cents per case for storage, two cents per case for handling, and one cent per case for clerical). Additionally, a multitude of peripheral services can be quoted on an individual basis: physical inventory count, freight company selection and routing, merchandise return and repair, and so on. In this way, the customer can choose the services desired and pay for each separately.

Attracting New Business

Various bundling strategies can be used to expand sales either by cross-selling – selling a new service to customers who have already bought an existing service – or by attracting new customers. In the cross-selling situation for An LSP warehouse, current customers (utilizing storage services) may be attracted to an innovative product-labeling service by the offer of a bundled price that results in a discount on the total cost of the two services. Bundling services in order to attract new customers can be efficient when the service attributes can be evaluated before purchase and when the core service is demand-elastic, that is, lower prices will generate higher volumes. Thus, non-customized services, where significant competition exists, would seem to be a fertile environment. Bundling equipment repair contracts with the rental of a fleet of lift trucks may attract new business customers for an LSP in the lift truck rental business.

In the computer service industry, manufacturers are finding that services formerly sold on an ad hoc basis can be sold more effectively if bundled. Many firms have been testing service bundles with customers in order to determine how customers want to buy the services. Clearly, the services, how they are combined, and how the bundle is priced affect the LSP's success in marketing the full range of its services.

Particularly salient to this discussion is the impact of new and innovative LSP services on the LSP's marketing strategy. Because a service is new and unique, the firm may encounter significant resistance in the market place until the new service has become popular. One way to overcome this resistance is to bundle the new service with existing services at a very attractive total price. This will reduce the risk of trial for the new service, and once the service has proven its value, future pricing schemes can reflect the value of the service. In addition, the new trials can be used to demonstrate the service to new customers who will then be expected to pay prices reflective of the service's value.

Promoting LSP Services

The promotional strategies for LSP services follow many of the same prescriptions as do those for most products. However, the unique characteristics of business services pose special challenges for the LSP marketer.

Communication with Employees

Personnel are vital to many people-based service businesses, and they can have a profound effect on the customer's satisfaction. Internal advertising to employees can promote an understanding of the firm's mission and customer service benefit and influences them on how the service is to be provided. Further, it is felt that internal advertising to employees motivates them to perform as well as defining management's expectations of them. Marketing communications directed to employees may be as important as those for potential customers. Such internal communications emphasize the company's purpose, its high standards of service, and the role that each employee assumes in creating satisfied customers.

Managers in LSPs should train and equip employees to carry out the service functions and also ensure that they have a clear grasp of the service product and process. A competent employee adds value to the service encounter and enhances customer satisfaction. In turn, service personnel play a critical role in reducing the risks that customers associate with purchasing intangible services, and that role can be enhanced by preparing personnel to offer information that anticipates customer concerns with the service. Some of the most effective LSPs invest significant time and dollars in hiring and training those employees who have the key interactions with customers. FedEx and UPS both have highly trained drivers who are skilled at dealing with both the firms at which they pick up products and the customer to whom the product is delivered.

Word of Mouth

Purchases of innovative and new LSP services are frequently considered to be riskier to many businesses than the purchase of products like components, repair parts, operating supplies and maintenance items. The reason is that it is more difficult for buyers to evaluate quality and value of a service than of a tangible product. As a result, buyers are more apt to be influenced by colleagues, peers, and other professionals who have already purchased and used the service. Promotion must concentrate on the dominant role of personal influence in the buying process and build on word-of-mouth communication. This can be done by persuading satisfied customers to inform others of their satisfaction and developing materials that customers can pass on to non-customers. It also helps to target opinion leaders in ad campaigns and to encourage potential customers to talk to existing customers. Some firms have arranged for non-customers to visit current customer operations to see the service/process in action, while others have found it useful to make presentations with current customers at professional conferences

LSPs can capitalize on the satisfaction of current customers and word-of-mouth promotion by featuring customers (and their comments) in advertising. For example, ads for warehouse management software can feature current customers explaining the virtues of a particular WMS product and its key salient features. CHEP, the pallet rental company, when it first entered the US market, focused on convincing one major consumer products firm in the US to use its system. That firm was Procter & Gamble, a company that is almost universally respected. Once Procter & Gamble adopted the CHEP system, many additional customers were acquired because of P&G's reputation for being ahead of the curve in adopting new business models. Testimonials and success stories on for new LSP services can always be highlighted on the LSP's website, further spreading the word on the key customer who are satisfied with a new service.

It is also useful for the LSP to induce a trade publication to feature an article about a new supply chain application developed by the LSP. For example, the website of *Global Logistics and Supply Chain Strategies* is called "Supply Chain Brain", and the website offers literally hundreds of case studies on interesting new solutions to difficult supply chain and logistics problems. An effective promotional approach for the LSP is to work with the Supply Chain Brain writers to develop an informative case study on the new service.

Developing Tangible Clues

Service marketers must concentrate either on featuring the "physical evidence" elements of their service or on making the intangible elements more tangible. "Physical evidence" plays an important role in creating the atmosphere and environment in which a service is bought or performed, and it influences the customer's perception of the service. Physical evidence is the tangible aspect of the service package, which the LSP can control. Attempts should be made to translate the image of intangible attributes of a service into something more concrete. IBM's Supply Chain Services division developed a clever set of television ads which depicted how IBM's processes and systems could be used to solve challenging supply chain problems. One advertisement featured a trucker confronting an IBM service representative sitting at a desk in the center of an US interstate highway. The trucker said he was lost, the IBM representative explained that through the use of IBM's logistics management systems and consulting, the trucker could easily find their destination. Here, the impact was to provide evidence of the effectiveness of the service and make the benefits tangible.

For LSPs, uniforms, logos, written contracts and guarantees, building appearance, and color schemes are some of the many ways to make their new services tangible. A handling equipment maintenance firm that provides free, written, quarterly inspections helps make its service more tangible. "Let Brown Do It", was a way for UPS to make its service more tangible. A spotless warehouse, with shiny floors, no litter, every product in place, is a way to tangibly show a prospective customer that a warehouse truly is "food grade" quality. A key concern for the LSP marketer is to develop a well-defined strategy for managing physical evidence – to enhance and differentiate service evidence through the creation of tangible clues.

3.5 Developing and Marketing New and Innovative LSP Services

The conventional process for developing new physical products – exploration, screening, business analysis, development, testing, and commercialization – appears to apply equally well to services. However, the design and introduction of new service offerings has been cited as one of the more pressing challenges for managers in the service sector:

"*New product development is inherently more difficult, messier and less successful in the service sector. If a service company perceives a new need and develops a new service, there is less confidence in the result because the service is not subject to the same rigor and predictable outcomes that new products are subject to in the R&D lab. Most service companies focus on geographic extensions of their service or on minor modifications rather than on truly innovative approaches. Innovation in the service sector is the result of trial and error. [...] Service firms have difficulty in linking innovations and imagination to execution of a new offer*" (Management Practice, 1984).

A major stumbling block to creating and launching a new service is "tangibilizing" the service concept. Traditional approaches, such as product prototyping, do not work with services because it is hard to prototype services that are often customized for individual buyers. However, Heskett offers five steps that a firm can take to improve the new service development process (see Figure 5). Consistent with the physical product innovation, it is important to create the proper organizational climate (for example, entrepreneurial culture, championing, taking risks).

Step		Description
1.	Establish a culture for entrepreneurship	Facilitate risk taking and new ideas by creating proper climate: providing R&D funds, doing customer-need research, allowing employees to voice contrary opinions.
2.	Create an organization to foster new service development	Assemble as 'cast': senior sponsor, who has authority; product champion, who provides continuity and enthusiasm; integrator, who brings the functions together and coordinates them; and referee, who establishes ground rules for the process and then administers them.
3.	Test ideas in the marketplace	New ideas must weather the acid of the marketplace because the service concept is intangible.
4.	Monitor results	Establish success measures and evaluate against these; track customer reaction.
5.	Reward risk takers	Reward those taking good risks, even when they are not consistently successful.

Figure 5: Steps for Enhancing the New Service Development Process (Source: Heskett, 1986)

A study of the process of developing new services tackled the following issues service development scenarios do managers in industrial service companies pursue, and what factors explain why some are likely to succeed while others fail? According to De Brentani, three scenarios describe the characteristics of successful situations, while two others characterize failed attempts at developing and launching new business services (De Brentani, 1995).

New Service Projects that Succeed

The three successful scenarios differ in the nature of the service initiative, the extent of service innovativeness, and the approach followed in developing and marketing the new service offering. The following profiles highlight key differences.

- The customized, expert service: These new services are relatively straightforward and inexpensive but are customized to the needs and operating systems of client firms. To respond to a customer's unique requirements, expert personnel are crucial to successful strategy execution. Examples include a customized learning center for a specific client offered by providers of supply chain software or warehouse management systems.

- The planned, pioneering venture: These are first-to-market services that are unique, complex, and expensive. A formal and carefully planned new service development process is a distinguishing feature of these service projects. Special attention is given to providing potential customers with tangible evidence that illuminates the benefits of the new service offering for them. Examples of these services include a terminal device linking LSP clients to multiple information origins and a computer-based remote access system developed by a trucking firm to locate and track shipments.
- The improved service experience: Represented here are equipment-based improvements made to a current service offering that increase the speed and reliability of the service process. Examples include an information systems–based expert production system provided by a computer systems organization and an order entry network developed by a large warehousing LSP.

All three of these scenarios share elements that are crucial to the success of new, innovative service offerings. Whereas business services can take many forms, new service success is closely associated with offerings that respond to market needs; that capitalize on a firm's reputation, skills, and resources; and that issue from a well-managed new service development process.

New Service Projects that Fail

What are the characteristics of new service initiatives that fail in the business market and what can we learn from them? Below are two scenarios for failure:

- The peripheral, low market potential service: These new services tend to be peripheral to the firm's core offerings, fail to bring added value to the customer, and enter a market with very limited potential. Failures of this type are common across service sectors in the business market. An LSP that decides to offer marketing research data about the client's customers would be an example of a service that is really peripheral to the typical LSP and for which the LSP has little inherent expertise.

- The poorly planned "industrialized" clone: These are complex new services that rely on "hard" technology (that is, equipment) for their production and delivery. Often, these are "me-too" services that offer no real customer benefits or improvements over those of well-established competitors.

Clearly, efforts to improve the efficiency and reliability of services by reducing customer contact and introducing equipment-intensive processes have succeeded in certain sectors of the business market. But, "such development efforts must be accompanied by apparent customer benefits that is, greater efficiencies and/or superior solutions to problems; a good fit with the capabilities of the developing firm; some competitive advantage in terms of service competitiveness; as well as a set of activities for researching, designing, and launching the new service offerings" (De Brentani, 1995).

3.6 Concluding Thoughts

Clearly, the development of a successful marketing approach for existing or innovative LSP services is a daunting task. LSPs have often shied away from marketing because they tend to focus on their operational excellence and core competencies. However, with the explosion of LSPs and the many opportunities in the marketplace, the future argues for more intensive efforts in the marketing realm. LSPs will have to learn how to craft marketing strategies that account for the unique and vexing characteristics associated with services as opposed to products. The LSPs that grasp the relevance of marketing and exhibit creativity in designing marketing strategies for their new service will be the ones to prosper in the future.

References

"Only 11% of global service employment could be practically outsourced overseas". Available: http://www.finfacts.com [Accessed 7 Oct. 2005].

Clemes, M.; Mollenkopf, D.; Burn, D. (2000). An Investigation of Marketing Problems Across Service Typologies, *Journal of Services Marketing*, 14 (6/7), 517-594.

De Brentani, U. (1995). New Industrial Service Development: Scenarios for Success and Failure, *Journal of Business Research*, 32 (2), 93-103.

Durvasula, S.; Lysonski, S.; Mehta, S.C. (2000). Business-to-Business Marketing: Service Recovery and Customer Satisfaction Issues with Ocean Shipping Lines, *European Journal of Marketing*, 34 (3/4), 433-352.

Global Employment Trends Brief (February, 2005). International Labor Organization.

Hertz, S.; Alfredsson, M. (2003). Strategic Development of Third Party Logistics Providers, *Industrial Marketing Management*, 32 (2), 139-149.

Heskett, J.L (1986). *Managing in the Service Economy*, Boston, MA: Harvard Business School Press.

Heskett, J.L.; Jones, T.O.; Loveman, G.W.; Sasser Jr., W.E.; Schlesinger, L.A. (1994). Putting the Service-Profit Chain to Work, *Harvard Business Review*, 72 (3), 164-174.

Jones, T.O.; Sasser Jr., W.E. (1995). Why Satisfied Customers Defect, *Harvard Business Review*, 73 (6), 88-99.

Management Practice (1984). Service Management, the Toughest Game in Town, 7 (2), 8.

Pitt, L.; Morris, M.H., Oosthuizen, P. (1996). Expectations of Service Quality as an Industrial Market Segmentation Variable, *Services Industries Journal*, 16 (1), 1-9.

Reichheld, F.F., (2001). *Loyalty Rules! How Today's Leaders Build Lasting Relationships*, Boston, MA: Harvard Business School Press.

Reichheld, F.F.; Sasser, Jr., W.E. (2001). Zero Defections: Quality Comes to Services, *Harvard Business Review*, 68 (5), 105-115.

Rust, R.T.; Zahorik, A.J.; Keiningham, T.L. (1995). Return on Quality (ROQ): Making Service Quality Financially Accountable, *Journal of Marketing*, 59 (4), 58-70.

Schiff, L. (2001). How Customer Satisfaction Improvement Works to Fuel Business Recovery at IBM, *Journal of Organizational Excellence*, 20 (2), 3-18.

Sharma, A.; Krishnan, R; Grewal, D. (2001). Value Creation in Markets: A Critical Area of Focus for Business-to-Business Marketers, *Industrial Marketing Management*, 30 (6), 391-402.

Zeithaml, V.A.; Bitner, M. J.; Gremler, D.D. (2006). *Services Marketing: Integrating Customer Focus Across the Firms*, Boston, MA: McGraw-Hill Irwin.

Zeithaml, V.A.; Parasuraman, A.: Berry, L.R. (1985). Problems and Strategies in Services Marketing, *Journal of Marketing*, 49 (2), 33-46.

Chapter 4

Innovation in Outsourced Logistics Services in Asia – The LSP Dilemma

Robert de Souza and Mark Goh

Abstract

While most shippers operating in Asia seek simple piecemeal freight and warehousing services, the logistics service provider (LSP) industry in Asia has been evolving from offering such traditional logistics services to offering a greater scope of albeit more complex supply chain solutions with an extended supply chain. The intent is to achieve higher service levels through an enhanced portfolio. This paper discusses the pressure to innovate against the backdrop of the shipper-LSP scenario in Asia, and suggests that the progress and value of innovative services in logistics outsourcing can be tracked via S-curves. The dilemma for the LSPs is to manage the existing logistics services cost-effectively and to provide an innovative array of suitable and timely integrated services, i.e., managing a portfolio of S-curves, preferably just ahead of the anticipated demand, of sufficient scale to justify the innovative investments against all the other demands that compete for the LSP's resources. A case study of an Asian LSP highlights the competitive dynamics that the S-curve framework can be used to analyze.

4 Innovation in Outsourced Logistics Services in Asia – The LSP Dilemma

4.1 Introduction

The service industry performs an increasingly dynamic and pivotal role in today's knowledge-based economies. Outsourced logistics services are no different. The logistics industry serves as a good example of the genesis and growth of an essential, new innovation-driven and value-recognized service-based industry, transformed from the traditional business notion of simple point-to-point transportation to that of serving the entire suite of customers' logistical needs and wants. Indeed, the quantum advances brought forth by science (plastic pallets), technology (RFID), and communications (GPS) in the new millennium have compelled logistics service providers (LSPs) to consider the potential of the new stream of soft resources, as exemplified by the prevalent technology, knowledge and relationship networks. These ingredients are necessary for both the shippers and LSPs to operate effectively and efficiently within the emerging business models of the 21st century. Moreover, they can help both parties to seize the opportunities, within the logistics environment and borrowed from beyond the logistics domain, to innovate and gain market leadership in a constantly turbulent environment, and to achieve faster time to value (popularly known as 'speed to market' in the logistics sense).

As competition becomes more intense and global, shippers (as we shall call the LSPs' customers) and LSPs invariably face many challenges, such as increasing demands, reduced costs, better logistics quality, better customer service, improved continuity of supply (Goebel et al., 2003), and faster time to value.

In Asia, manufacturing organizations are constantly compelled to relentlessly restructure and re-engineer to increase their effectiveness and satisfy or exceed customer expectations. This realization obviously necessitates that such firms look beyond their traditional organizational boundaries, rely more on their chain partners (e.g. in logistics), to evaluate how the resources and capabilities of service providers and shippers can be better exploited to create exceptional value. These firms rely on erstwhile smart strategies to develop and sustain competitive advantage. In a supply chain context, two strategy approaches come to mind.

First, these firms depend on innovation through either disruptive technologies, customer discovery, competitive benchmarking, or even deep process improvements to achieve that needed commercial advantage. In this regard, Chapman et al. (2003) have studied some of the factors that nurture innovations in logistics services and attempted to show how the new soft resources can contribute to better innovations within the realm of logistics. Second, premised by outsourcing to reduce cost and focus on core competencies, LSPs are obliged to lend shippers a hand in executing that desired innovation. However, are the LSPs innovative in the provision of services or are they not doing enough to meet the challenges and needs of the shipper community? Both views prevail in the supply chain community in Asia today, without any clear conclusion and direction.

Nevertheless, in the present climate, what is driving innovation in logistics services is most probably the effort to extract greater value from innovations in logistics outsourcing beyond the provision of the traditional commodity services such as warehousing and transport. While most Multi-National Corporations (MNCs) operating in Asia may seek simple piecemeal freight and warehousing services, the LSP landscape in Asia is gradually offering greater scope, and richer but more complex supply chain solutions, involving an extended supply chain of suppliers, customers, and other players.

This perceived gap between the needs of the shippers and the LSPs' capabilities has led to an interesting development. That is, one where the LSPs may possess capabilities that have not yet been fully exploited, and one where shippers are still not fully convinced of the potential of a greater degree of outsourcing of higher value logistics services. Some of the higher value logistics services in Asia include demand forecasting, information management, supplier management, and call centre management. Hence, selecting fewer but better LSPs, and having them commit to bigger and longer term service contracts may just reap those elusive advantages of scope and scale that prevent them from justifying outsourcing (beyond warehousing and transport) in the first place.

This naturally puts the LSP industry in the dilemma of having to create greater genuine value in outsourced logistics services while containing costs. Given that there are many different interpretations of innovation, e.g. Flint et al. (2005), for this paper, we treat innovation as a subjective novelty which is open to a conscious management process and which aims at economic exploitation. Also, as there are numerous forms of innovation, we will concentrate on product/service innovation and disregard process innovation, and business model innovation. Thus, by outsourced logistics innovation, we mean any kind of logistics-related service that is new and helpful to a particular focal audience (a company, client industry or other). This paper suggests that as companies migrate from product innovation to solution (service) innovation (Shepherd and Ahmed, 2000), the LSP can likewise build greater value in logistics outsourcing through deliberately prescribed service innovations, thus leading to better firm performance for the shippers.

4.2 S-curves and Logistics Services Innovation

While the extant literature is replete with theories on how industries change or product and service offerings evolve, some simple yet elegant models have become household features. For instance, Porter's (1980) five-forces framework (potential entrants, buyers, substitutes, suppliers, and industry competitors) is a static snapshot that is a useful way of assessing the viability of a whole industry to innovation investments. Complementing Porter's framework, McGahan (2000) introduces a framework that identifies four models of industry evolution Each model has distinctive characteristics in terms of industry structure, historical innovation paths, investment opportunities, and performance drivers. The S-curve framework is yet another framework for analysing competitive or market dynamics (Christensen et al., 2004). Christensen et al. have succinctly presented how companies ought to be able to recognize the signals for change, how to identify and develop new innovations to supplant the existing ones. Further, they highlighted that new disruptive innovations has the greatest potential for long term industry change but such innovations are difficult to identify.

As a brief digression, we now discuss the nuances of the S-curve. The lower end of the S-curve represents a *fluid* state where technologies, offerings and markets are ill-defined. It is here that the ability to predict or anticipate new demands needs a new science. The business processes involved at this stage are loosely held and the organizations of the practices are essentially ad-hoc. Stability of operations is uncertain. There is much trial and error and frequent changes to strategy. A steeper middle part of the S-curve corresponds to the *transition* state where a service performance-maximizing opportunity exists after the dominance of a particular service offering such as warehousing or transport takes hold. At this stage, we can assume that the visibility of market demand is clearer and the value of the dominant offering or innovation is established. The innovations are now focused on maximizing performance. The upper part of the S-curve denotes the cost-minimizing *specific* state when the service offering may be reaching the upper limit or maturation level of its capability and begins to assume a commodity-like form. In this stage, most of the industry's competitors would have followed responded accordingly, either through imitation or new innovations. Processes and practices are more rigidly aligned with little margin for profit. Further similar innovations, if at all possible, are focused on minimization. The resistance to substantive innovation is relatively stronger.

Since the 1960's, innovators have been familiar with S-curves (Martin, 1994). In order not to be sucked into the vortex of a permanent specific state, Martin was probably the first to suggest that management had a responsibility to nurture a continuous succession of new products through sustained technological innovation and that the innovation process must be managed properly to produce commercially successful product winners. Today, it still provides a basis for the understanding of logistics services innovation. In our case, one aspect could be the evolution of technology-based logistics and supply chain management capabilities. On technology-based logistics, Wu (2006) has asserted that shippers will rely on the LSPs, beyond distribution services for more technological innovation as the latter successfully expand capabilities into these areas, and provide end-to-end supply chain integration and visibility.

Though the S-curve framework focuses on how industries evolve, it ignores how LSPs may transcend product generations (i.e., across a series of S-curves). For ease of illustration, we have selected the S-curve model as the simplest and most representative but we note its constraints and the other models that could be applied for deeper analysis.

In applying the three states of the S-curve to the logistics landscape, we can draw some parallels. If we scrutinize what is commonly being outsourced globally, a recent 3PL survey (Langley, 2005) clearly shows that the traditional activities of outbound transportation and warehousing are still frontrunners (see Figure 1). At the same time, the higher value added services such as inventory management and lead logistics provider (LLP) or 4PL services have the potential to develop in outsourcing maturity. Hence, this begs the next question: have these lesser-valued outsourced services reached the *specific* state of the S-Curve? And if so, after what period in time? Today, it may well appear to be so in Asia, where the provision of transportation and warehousing services is increasingly perceived or accepted as commodity or mere entry-level services for an LSP-MNC relationship. However, if we consider the provision of innovative services in the areas of life sciences logistics and exhibition logistics, the verdict may well be reversed. Here, the extrapolations of the S-curve might be different - one may follow a shallower growth path to a more modest upper limit and the other might grow more aggressively to a much higher limit in a short time. There is yet another corollary: when will the LSP move up the S-curve of the higher value added services as indicated by Figure 1, in time for the LSP to ride on it as an innovative service offering? Timing is the key pre-determinant.

+ Common

- Outbound transportation (WE 88%, AP 96%, LATAM 84%, NA 78%)
- Warehousing (WE 72%, AP 88%, LATAM 55%, NA 63%)
- Customs brokerage (WE 49%, AP 67%, LATAM 55%, NA 63%)
- Inbound transportation (WE 66%, AP 68%, LATAM 68%, NA 58%)
- Freight forwarding
- Transportation management
- Freight bill auditing/ payment (very big in North America)

- Rate negotiation
- LLP/ 4PL services
- Material management
- Inventory ownership
- Order entry/ processing/ customer service
- Customer and supplier compliance
- Factoring (trade financing)

- Common

Figure 1: Commonly Outsourced Logistics Services (Source: Langley, 2005)

Can we, therefore, use the S-curve model to predict at which point during the *transition* phase an innovative logistics service offering should be spun off to counter (or balance) the cost-minimizing state? With sufficient hindsight and event tracking, this is theoretically possible though highly consuming in time and capital-resources. However, skipping to a future trajectory is difficult because of the market opportunity limits and because the steepness of the slope is difficult to anticipate. Accurately forecasting customer needs requires adept technical, management, political, economic and other external environmental scans, competitor analysis and deep market insights. Observing these subtle signals of change can help to determine the inflexion point for a call to action. In short, we can then choose either to depart from an existing curve or start a new higher performance level service innovation while still retaining the other services in the existing portfolio. This raises the barriers to entry and maintains the gradient of the trajectory. Put simply, the progression of each successive (substitution or new) S-curve (labeled as X_0 ,..., X_3 in Figure 2) leads to a straight line path of sustainable logistics services innovations and the continued relevance of the LSP's optimally matched services to the needs of the shippers.

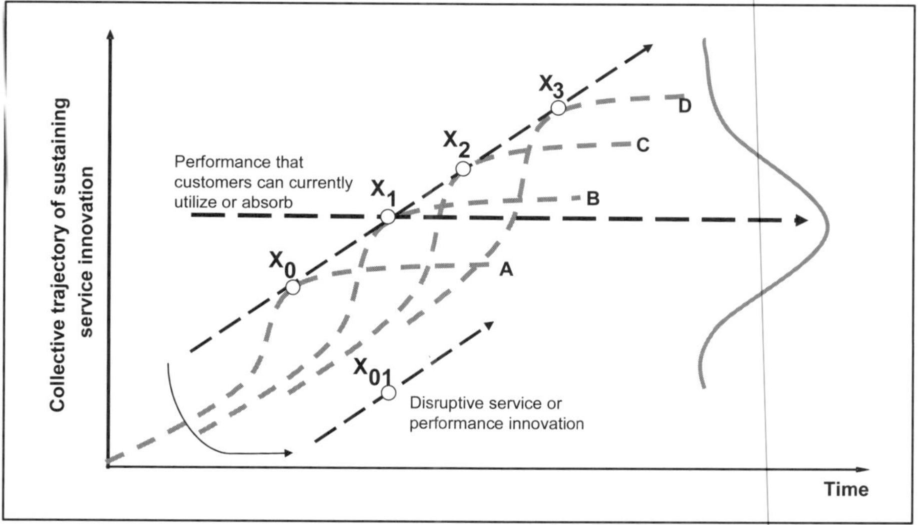

Figure 2: Understanding the Dynamics of Innovation (Source: Christensen et al. 2004; Martin, 1994)

Therefore, one would expect the evolution of logistics services innovation to follow the envelope denoted by points X_0 ,..., X_3, which should represent the different jump states and bring the LSP to the next level of logistics services sophistication. Indeed, capturing and manipulating the envelope formed by the discrete quantum leaps can sometimes make or break an LSP desiring to grow beyond the local market operation as such "jumps" in innovation hold significant geographical and operational implications. The challenge as always is to sustain the momentum of innovation.

4.3 Staying Ahead of the Innovation Game

Figure 2 illustrates the complexity of understanding the dynamics of sustaining innovation, especially for services, which can be commoditized rather quickly. Focusing on sustaining service innovations without due regard to the range of services or service performance that the shippers can comprehensively utilize or outsource can result in poor returns on investment and unrealized returns. From an investment or cash flow perspective, the S-curve begins with a negative outlay at the start, pursues aggressive rapid returns, followed by slower or more stable (mature) returns, and then a progressive decline in earnings. Therein lies the dilemma for the LSP. Investing too early in a disruptive service innovation may not net the requisite customer base for the LSP, particularly in technology- or infrastructure-based logistics services.

Exceeding the performance level that a significant segment of the shipper's market can absorb opens the door to a "disruptor" (LSP) who enters and attacks a given (lower) tier of the market distribution. This is already attested by the numerous LSPs in Asia providing warehousing and transportation services at or near commodity prices. That makes it difficult to elicit higher profit margins for incumbent LSPs, without further defensive investment commitment in innovation. This relegates the LSP to an even lower level of market dynamics, focusing solely on cost to compete and survive.

When a low-cost competitor with a lower cost base enters this competitive space during the transition phase of a particular S-curve leading to cost minimization strategies at the *specific* phase, it will enjoy a cost advantage only as long as the higher cost competitor chooses to remain in the market. The higher cost competitor must then actively seek an innovative trajectory shifted to the lower right of the original trajectory, labelled in Figure 2 as X_{01}, merge its services, be acquired or let its services decline and/or disappear. Similarly, an LSP that continues to seek or create competitive advantage through a strategy of differentiated service innovations and higher service levels must constantly seek to match their innovations with shippers that value such points of differentiation. Thus, bigger LSPs, for better or worse, have the challenges of opening new markets and simultaneously defending their differentiation during that (extended) window of opportunity. Christensen et al. (2004) describe how organizations might respond to such challenges from the Resources, Processes and Values (RPV) perspective, which suggests that "*resources (what a firm has), processes (how a firm does its work) and values (what a firm wants to do) collectively define an organization's strengths and weaknesses in driving innovation.*"

In response to the S-curve logic, some LSPs have elevated themselves from being just tactical LSPs to being supply chain integrators. However, Langley (2005) rightly highlights (see Figure 3) that the market globally (including Asia) still requires the former, and those LSPs providing inventory management and LLP/ 4PL services are only slightly ahead of the demand curve. Clearly, the key to a successful shipper-LSP relationship is whether the shipper's expectations are matched to the appropriate LSP business model and whether the LSP can provide the best business value at that point in time. In the Asia-Pacific region, for example, 67 percent of shippers still view the LSP as a tactical service provider, compared to 30 percent of shippers who view the LSP as a supply chain integrator. Yet, LSPs, in anticipation, continue to focus on and innovate in the role of an integrator, sometimes at their own expense and thereby risk a loss of focus. This begets the next question: where and how the emerging LSP in Asia should position itself.

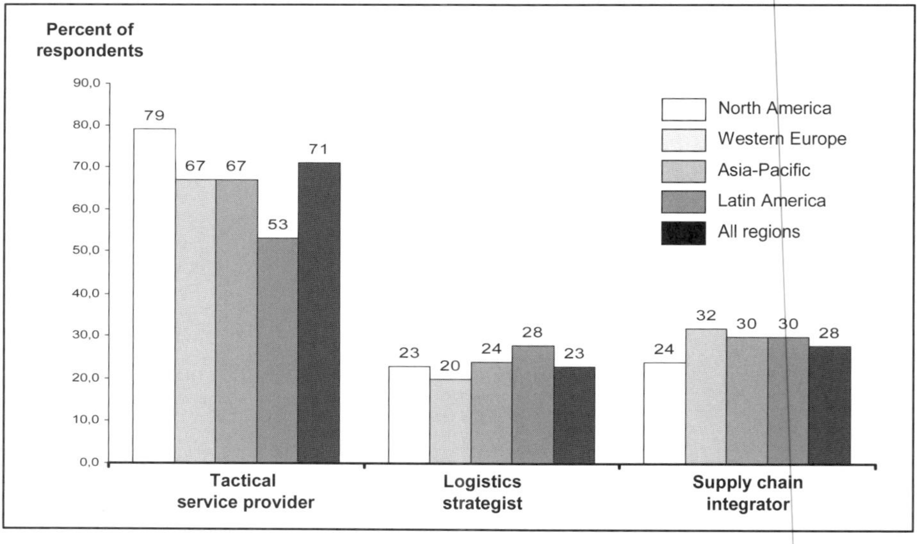

Figure 3: LSP Roles (Source: Langley, 2005)

4.4 Innovation Journey of an Asian LSP

We now provide a case study of how an Asian LSP has grown with shippers in the region due to its ability to read the logistics scene, and has responded innovatively to meet and sometimes pre-empt the growing needs of shippers, despite the dilemma described earlier in this paper.

The YCH Group, an LSP based in Singapore since the 1970s, is one of the pioneering Asian LSPs that entered China more than a decade ago, by servicing its network of MNCs such as Dell, Motorola, Canon, and Ciba Specialty Chemicals when these companies decided to relocate some of their manufacturing operations to China. Today, YCH has a logistics presence and operations in Chinese cities such as Shanghai, Tianjin, Hangzhou, Xiamen, Shenzhen, and Hong Kong. As part of its continued effort in sustaining infrastructure-based logistics services innovation, YCH has agreed with the Shanghai Lingang International Logistics Development Company, to develop a 4.3-hectare distribution park in the city's free trade port, to jointly house a world-class dangerous goods (DG) storage facility within the Lingang Logistics Park meeting the growing demands of the newly-emerging container port located at Yangshan (Yang, 2006).

This investment project augments YCH's already-established logistics network in Shanghai, in line with its strategic plan to continuously and conscientiously develop pervasive supply chain capabilities in China and innovatively manage end-to-end supply chains for YCH's MNC customers. Through this strategic location decision, a performance-related innovation is undertaken whereby YCH is now in a stronger position (located at the deepest and first port of entry to central China) to manage the passage of raw materials and sub-finished goods into China, and using this storage location to forward feed to various manufacturing facilities in China, and then onto other hubs in Asia. At the same time, being located in a free trade processing area will also assist in managing the spares and returns of products for the shippers, as well as faster distribution of finished goods into the domestic Chinese market or internationally. This novel choice of location, while more expensive than the typical manufacturing zones elsewhere in Shanghai, is instrumental in offering a three-pronged advantage of being a free trade zone, an export processing zone, and a bonded logistics park.

YCH can now rapidly implement sophisticated logistics service programs. Two leading programs provided by YCH are vendor managed inventory and innovative supply chain management solutions. In the case of the latter program, YCH's virtual hub solution involves the provisioning of value-added services desired by MNCs operating in Asia: postponement, configuration, merging, and repackaging.

With the planned DG facility in the Lingang Logistics Park, YCH is well placed to meet the growing demand for DG storage services in Shanghai and eastern China, which is already home to many large bulk and specialty chemicals companies, and petrochemical companies such as BASF and Ciba. Indeed, this investment is an innovative step to move in sync with the client companies as YCH has been in Shanghai for the past ten years, serving world-class fine and specialty chemical MNCs such as Ciba Specialty Chemicals and DSM. To quote the Chairman of the company, Dr. Robert Yap, "*as a key player in the specialty chemicals supply chain, this development is a natural and necessary progression to enhance (innovate) our existing capacities. It is also to further expand and strengthen our network into the greater Shanghai region in which we have already built a strong base and foundation*" (Yang, 2006).

Aside from this, YCH is also extremely active in the city of Tianjin (located in Northern China) which serves as the North China regional hub for MNC production and hubbing operations. Having developed a 3.3-hectare Tianjin distribution park in 2004 and a 5.2-hectare airport logistics park in 2007, YCH has witnessed remarkable growth in Tianjin. YCH is already poised to leverage its gateway position in Tianjin to better serve its existing and potential customers around Northern China. Presently, YCH is the top importer in Tianjin, having handled US$3.3 billion of imports in 2005, and helps to distribute final goods and components to neighboring countries in North Asia such as Japan and South Korea. YCH is currently helping Motorola in Tianjin to manage the supply chain of one of Motorola's largest manufacturing bases, by providing the Virtual Hub™, a proprietary, award-winning, innovative solution which facilitates the virtual sourcing of raw materials for Motorola "to buy anywhere, make anywhere, and sell anywhere" (see *http://www.ych.com/press_release_0706.htm*).

With the presence of these strong and multi-capability logistics nodes in China, and other logistics locations in ASEAN and South Asia (India, Australia, and New Zealand), YCH is in a prime position to exploit emerging country locations, and be the strategic bridge and enabler for its current and potential manufacturing customers (including exporters and importers), to connect their supply chain throughout the Asia Pacific region. Having risen through the rungs of the value-added pyramid, from transportation, warehousing, and inventory management to assembly and testing, YCH has developed in-depth expertise in every aspect of supply chain management, from the very core of operational execution to that of leveraging technology and knowledge-based activities. Clearly, this is a positive outcome of continuous innovation, albeit it necessitates that YCH maintains the momentum of innovation investment.

In terms of technological innovation, YCH is a pioneer in leveraging technology to enable and integrate the supply chain of its customers. According to Lin (2007), logistics technologies are classified into four types: data acquisition technologies, information technologies, warehousing technologies, and transportation technologies. Lin also reports that the innovation in logistics technologies in China is significantly influenced by organizational support, the quality of the human resources, environmental uncertainty, and governmental incentives for the LSPs. In this regard, YCH has progressively employed innovation in each of the logistics technologies for shippers, and has deliberately attempted to stay on the envelope of the innovation S-curves, as defined by Figure 2.

The same applies for YCH's infrastructure-based innovation from general storage facilities to DG facilities, and onto processing parks. This is necessary as the logistics industry in Asia is still highly fragmented and complex (Yang, 2006) (especially for emerging economies), and there are many breeds of players in various segments of the logistics landscape. The unique value propositions of these factors of technology and logistics parks, as provided by YCH, focus on a niche that is difficult to replicate, and finds very few competitors in the same league of LSPs. By emphasizing the integration of the physical, information and financial flows throughout the supply chain, YCH recognizes that innovation will remain the primary differentiator and driver of growth and expansion. Indeed, the slew of recognition accorded by the Singapore and Chinese governments belies YCH's continuing effort and re-tooling to meet the changing requirements of the customer.

Today, the YCH Group's suite of innovative award-winning end-to-end supply chain management solutions include Intribution™, Intrabution™, Retrogistics™, Y-Merge™, and V-Hub™. None of these innovations are rare. Rather, they employ leading-edge web-technologies across industry clusters, ranging from consumer and electronics to chemical and healthcare, to deliver powerful competitive advantages for shippers. A strong proponent of innovation, YCH has already been recognized for its 7PL™ approach in seamlessly integrating supply chain strategy with execution in the logistics industry. In 2005, YCH won the prestigious Singapore Innovation Class (see *http://www.ych.com*).

In doing so, the S-curve for YCH has translated up the innovation ladder from Curve A to Curve D, through a series of strategically poised innovations adopted over time. Figure 4 presents YCH's innovation journey in relation to the S-curve structure presented in Figure 2. As observed, YCH has stayed on the envelope of a series of strategically-timed and deliberately maneuvered innovations, ranging from smarter transport arrangements, to VMI, setting up logistics parks (as integrated logistics provider), and logistics IT service provider (e.g., 3-D bar codes and RFID).

According to Christensen et al. (2004), YCH has aggressively tried to identify new disruptive innovations, whether in logistics infrastructure or technology, and YCH has leveraged those innovations to invoke long-term industry change for the LSPs in Asia. In so doing, YCH has ridden the crest of a series of S-curves and applied Martin's (1994) thinking to deliver new logistics services at reasonable cost and investments, and managed the innovation process to produce the right order winners.

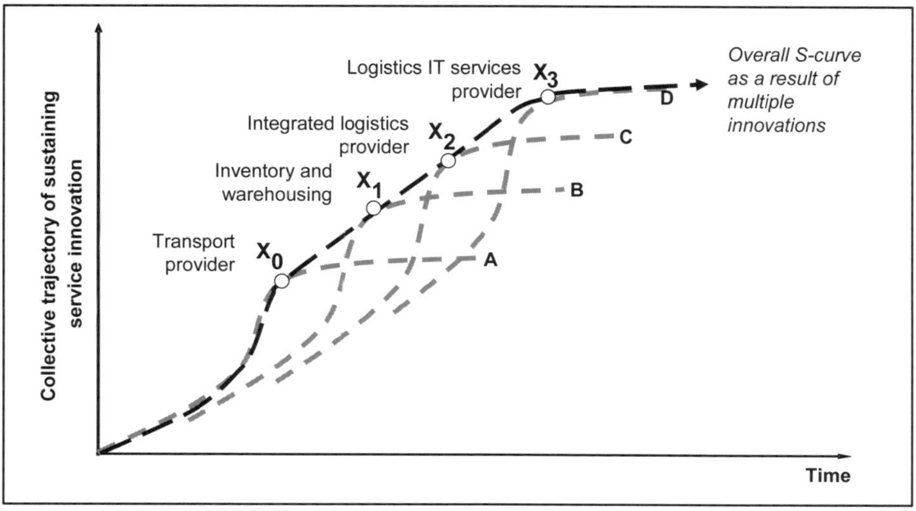

Figure 4: Evolution of YCH's Logistics Services Innovation

4.5 Search for the Eternal Sweet Spot

Understanding the complex dynamics of innovation, winning the competitive battles and defending the high ground have become part and parcel of the business of LSPs as their industry consolidates. At the same time, it has become more attractive for the shippers to select (these) fewer and longer term, regional and global LSP partners. The battle, based on the innovation model described in Figure 2, is for the middle ground (i.e., the shippers that tactically outsource but retain the higher value functions of planning and inventory management). To win this non-asset-based services ground, an IBM study (Bitner and Moore, 2006) has suggested that this requires the emergence of high process conformance providers who can provide not only a deeper level of operational effectiveness but who can also demonstrate strong efficiency and continuous process improvements, process simplifications, strict adherence to global standards, and mitigation of complexity innovations.

Current evidence is still inconclusive; a recent survey (Langley, 2005) has indicated that although price is still the deciding factor, other factors that follow such as product, service, experience, and access (Figure 5) are fast eroding this perception. The sustainable competitive advantage for any LSP is driven by constant innovation dependent on satisfying constantly changing customer fashions, trends, tastes and patterns of demand, a pattern which the case of YCH is very much akin to. In such an environment, the degree of and willingness to pursue innovation depends on the close interaction between the LSPs and their customers, and this relationship includes rapid and consistent information exchange as well as the control of the actual physical supply chain (Mowatt, 2006).

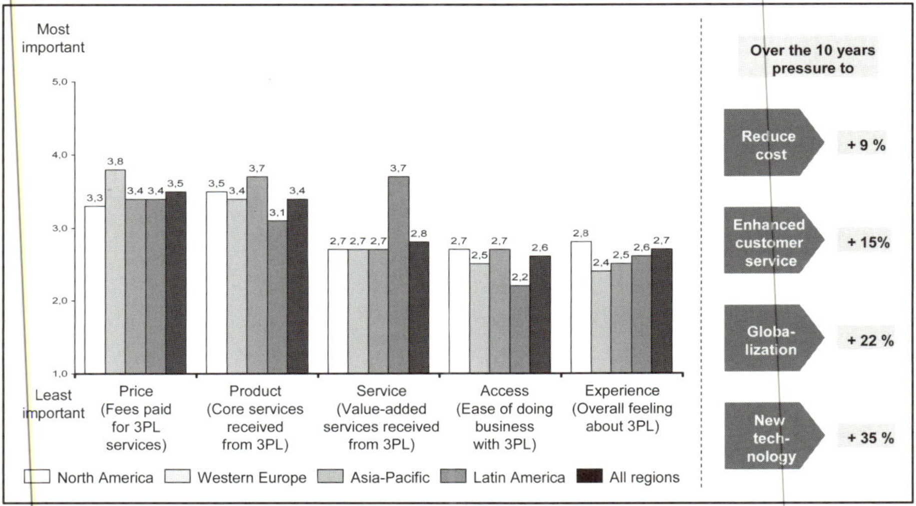

Figure 5: Ranking of Key LSP Selection Attributes (Source: Langley, 2005)

4.6 Conclusion

This paper is the subject of ongoing research at the Logistics Institute – Asia Pacific. We have discussed the pressure to innovate with the backdrop of the shipper-LSP scenario in Asia, using the case of YCH. Figure 2 has illustrated how the progress and value of innovative services in logistics outsourcing can be tracked via S-curves. We have mapped the case of YCH's innovation journey against that of Figure 2, and presented Figure 4 to show how LSPs in Asia have leveraged on innovations as a key service offering.

However, as Langley (2005) shows, the pressure to innovate has gone hand-in-hand with the pressures to contain costs (possibly leading to an early maturation of such services), to achieve higher service levels through an enhanced services portfolio, and the adoption and deployment of new technologies. Herein lies the dilemma for LSPs. The challenge is not only in cost-effectively managing each logistics service but also in providing an innovative array of suitable and timely integrated services (i.e., managing a portfolio of S-curves preferably just ahead of the anticipated demand of sufficient scale to justify innovative investments presently in the face of all the other demands that compete for the LSP's resources). Adopting a me-too business model or a lack of prioritization of innovative service opportunities (see Christensen's RPV Theory) may instead create poorer than anticipated returns and open this market space to further disruptions and greater consolidation among the LSPs.

References

Bitner, K.; Moore, D. (2006). *Building value in logistics outsourcing: the future of the logistics provider industry*, IBM Institute for Business Value. Available: http://www.1.ibm.com/services/us/index.wss/ibvstudy/imc [Accessed 16 Mar. 2006].

Chapman, R.L.: Soosay, C.; Kandampully, J. (2003). Innovation in logistic services and the new business model: A conceptual framework, *International Journal of Physical Distribution & Logistics Management*, 33 (7), 630-650.

Christensen, C.M.; Anthony S.D.; Roth, E.A. (2004). *Seeing what's next: using the theories of innovation to predict industry change*, Boston, MA: Harvard Business School Press.

Flint, D.J.: Larsson, E.; Gammelgaard, B.; Mentzer, J.T. (2005). Logistics innovation: a customer value-oriented social process, *Journal of Business Logistics*, 26 (1), 113-147.

Goebel, D.L.; Marshall, G.W.; Locander, W.B. (2003). Enhancing purchasing strategic reputation: evidence and recommendations for future research, *Journal of Supply Chain Management*, 39 (2), 4-13.

Langley Jr., C.J. (2005). *2005 Third-party logistics – results and findings of the 10th annual study*, unpublished manuscript, The Logistics Institute (TLI).

Lin, C.-Y. (2007). Factors affecting innovation in logistics technologies for logistics service providers in China, *Journal of Technology Management in China*, 2 (1), 22-37.

Martin, M.J.C. (1994). *Managing innovation and entrepreneurship in technology based firms*, New York: John Wiley.

McGahan, A.M. (2000). How industries evolve, *Business Strategy Review*, 11 (3), 1-16.

Mowatt, S. (2006). New perspectives on the supply chain and consumer-driven innovation, *International Journal of Services Technology and Management*, 7 (13), 515-534.

Porter, M. (1980). *Competitive strategy techniques for analyzing industries and competitors*, New York: Free Press.

Shepherd, C.; Ahmed, P.K. (2000). From product innovation to solutions innovations: A new paradigm for competitive advantage, *European Journal of Innovation Management*, 3 (2), 100-106.

Wu, Y.-C.J. (2006). Assessment of technological innovations in patenting for 3rd party logistics providers, *Journal of Enterprise Information Management*, 19 (5), 504-524.

YCH Group. Available: http://www.ych.com [Accessed: 1 Aug. 2007].

Yang, P. (2006). Innovation key to YCH's expansion, Shanghai Daily, 25 September 2006. Available: http://www.shanghaidaily.com [Accessed: 1 Aug. 2007].

Chapter 5

A Focused Innovation Model for Logistics Service Providers

A. Dale Flowers, Stanton G. Cort,
Les B. Artman and Ronald H. Ballou

Abstract

This paper is targeted to leaders of logistics service providers (LSPs). It makes several key contributions to the body of knowledge about managing and leading such firms. First, it introduces the goal of becoming the "essential link" in the markets that an LSP serves. The balance of the article provides the concepts necessary to achieve this goal. For the first time in the literature, we present an integrated innovation model representing the union of the Stage-Gate™ System and open innovation. We relate these specifically to LSPs. Once we understand the new model of innovation, to what do we apply it? We present three operating models of market leaders, and relate these models to LSPs. Since this combination of concepts has not appeared in the literature on LSPs, we review recent LSP publications to illustrate how well the concepts do apply. Finally, we provide a summary of how the market leader concept can help an LSP to focus its innovation efforts, using the integrated innovation model, to achieve the goal of being the essential link for the markets that it serves.

5 A Focused Innovation Model for Logistics Service Providers

5.1 Introduction and Background

For several years it has been common for logistics service providers (LSPs) to offer an ever greater variety of services to their customers. In explaining the value proposition of LSPs (specifically, third party logistics providers), Atkinson (2007, p. 20) notes:

"*Gone are the days when companies simply asked their third-party logistics providers (3PLs) to do the 'pallet in – pallet out' drill. More companies expect 3PLs to perform a host of complex, intricate and even strategic tasks that were once performed by the companies themselves, with never a thought that such activities, being so critical to the company's success, could ever be outsourced.*

In many cases the arrangements are working out so well that not only are companies asking their 3PLs to handle more and more of these value-added tasks, but the 3PLs themselves are coming up with innovative value-adding ideas that allow companies (their customers) to focus more on their core competencies."

However, Hertz and Alfredsson (2003, p. 148) conclude: "*Moving into the emerging TPL (Third Party Logistics) business the development seems, to a large extent, to have been based on following and integrating with existing customers rather than being a strategic choice.*"

Lieb (2005, p. 24) adds: "*Pressures from their (3PL companies) boards to grow revenues as quickly as possible led to the perception that any potential customer was a good one. Consequently, some providers attempted to be everything to everyone – clearly an unworkable strategy.*"

So it appears that LSPs have not been taking their own advice: to focus on their core competencies. The strategic questions are: How can core competencies be developed and what should they actually be? It is our purpose to provide concepts that LSPs can use to answer these important questions.

In the next section, we will set the stage for the balance of the paper by introducing the goal of the LSP to be an "essential link" within the supply chains that it serves. LSPs need to achieve this goal and sustain the performance over time in order to be truly successful.

In order to become an essential link, LSPs have to innovate. In section 3, we will present an integrated model of innovation that employs the union of ideas from two models in prevalent use in manufacturing companies. These two models have proven track records and, as we will see, the resulting model fits LSPs and their service offerings. This integrated innovation model answers the first of the two questions listed above; how should the LSP develop its core competencies?

To answer the second question of what the core competencies should be, we will introduce three different models for market leaders in section 4, and relate the models to LSPs. These models come from *The Discipline of Market Leaders* by Treacy and Wiersema (1995). These concepts help provide the focus for the innovation process by helping LSPs define the type of company they want to be.

In section 5, we illustrate how well the market leader concepts apply to LSPs by using recent LSP literature.

Our concluding section 6 summarizes the focused innovation process that results from the combination of our ideas. It guides LSPs to develop their core competencies, (the integrated innovation process) and what they should be (the "focus") so as to become the "essential link" between their customers and suppliers.

5.2 Becoming the Essential Link

As those who have tried have learned, becoming a successful LSP is very challenging. On one side, customers place more and greater demands on the LSP, while seeking to diminish the power of the LSP over the customer. In some cases, this can mean trying to acquire the competencies that the LSP possesses in order to limit reliance on the LSP. On the other side, the suppliers/intermediaries for an LSP may look to expand their scope of activity to include functions that the LSP is performing for its customers. The supplier of vehicle routing software to an LSP may try to sell the software directly to the LSP's customers. So what is the LSP to do?

The LSP needs to map a strategic and tactical course that allows them to achieve and sustain control of their relationship with both their customers and their suppliers/intermediaries. This control allows for the LSP to remain autonomous in its decision making and to maximize its ability to develop its strengths. The development of those strengths should help it to sustain its competitive advantage and ultimately achieve commercial success. Yet, the LSP needs to have both its customers and its suppliers/intermediaries do business with the LSP in order to deliver value to the customers.

However, as we noted above, both the customers and the suppliers/intermediaries to an LSP could be tempted to circumvent/disintermediate the LSP's role in the value chain. The LSP needs to prevent this from occurring by causing both the customers and suppliers/intermediaries to conclude that achieving their own commercial success is dependent on working with the LSP. At this point, the temptation to circumvent/disintermediate is weaker than the return on investing in the relationship with the LSP. The LSP has thus become the essential link between its customers and its suppliers/intermediaries.

In order to become and remain the essential link, the LSP must continually and swiftly innovate in a variety of ways. Otherwise, they may lose business to their own customers, who acquire the knowledge and skills from the LSP over time, or to suppliers/intermediaries who expand their offerings in the hope of obtaining a larger proportion of the business of the LSP's customers. Manufacturing companies have been facing such challenges for some time as the rate of new product introduction has accelerated. In the next section, we present a new, integrated innovation model representing the union of the two dominant systems in use by manufacturing companies.

5.3 An Integrated Innovation Model

Two popular models of innovation are in widespread use. The first and older of the two is the Stage-Gate™ System popularized by Cooper (2001). The essence of this system is to have a structured, disciplined system for the development of new products and services that is based on a project passing through each of five stages or gates. Figure 2 contains a suggested Stage-Gate™ innovation process. Cooper (2001) provides extensive explanations for each stage and gate in the process.

Figure 1 lists a sample of companies reported by Cooper to use the Stage-Gate™ system.

3M	Dupont	IBM	Procter & Gamble
Air Products	Emerson Electric	ICI	Reckitt-Benckiser
American Express	GE-Honeywell	Lego	Rohm and Haas
Asahi Chemical	General Mills	Lucent	Royal Bank of Canada
B. F. Goodrich	Guiness	Microsoft	SC Johnson Wax
Bayer	Exxon Chemical	Mobile Chemical	Toray Chemical
Carlsberg	Hewlett-Packard	Nortel Networks	Unilever
Corning Glass	Hillshire Farm	NYNEX	VISA
Dow Chemical		Pillsbury	

Figure 1: Sample Companies Using the Stage-GateTM Process

The second model comes from Chesbrough (2003) and is called "open innovation." The essence of the open innovation model is that a company's boundaries need to be very porous as concerns the innovation process. He contrasts it with systems that he calls "closed innovation," into which he places the Stage-Gate™ system of Cooper. His characterization of closed innovation systems includes (Chesbrough, 2003, p. xxvi):

- "*The smart people in our field work for us.*"
- "*To profit from R&D, we must discover it, develop it, and ship it ourselves.*"
- "*If we discover it ourselves, we will get to the market first.*"
- "*The company that gets an innovation to market first will win.*"
- "*If we create the most and the best ideas in the industry, we will win.*"

- *"We should control our IP, so that our competitors don't profit from our ideas."*

He contrasts these principles one-for-one with his concepts for open innovation (Chesbrough, 2003, p. xxvi):

- *"Not all the smart people work for us. We need to work with smart people inside and outside our company."*
- *"External R&D can create significant value; internal R&D is needed to claim some portion of that value."*
- *"We don't have to originate the research to profit from it."*
- *"Building a better business model is better than getting to market first."*
- *"If we make the best use of internal and external ideas, we will win."*
- *"We should profit from others' use of our IP, and we should buy others' IP whenever it advances our own business model."*

Chesbrough (2003, p. 64-65) emphasizes the dominance of the business model for innovation, and lists the functions that the business model should perform:

- "*To articulate the value proposition*"
- "*To identify a market segment*"
- "*To define the structure of the firm's value chain*"
- To specify the estimated revenue, cost, and margins
- "*To describe the position of the firm within the value network*"
- "*To formulate the competitive strategy*" for differential advantage

In spite of the contrast that Chesbrough claims, it is possible to integrate the concepts from the two models and include the best of both. For example, there is an overlap between Figure 2 and the business model functions of Chesbrough. A careful review of the details for the Cooper process reveals that all the business model functions listed above are included in the Stage-Gate™ innovation process. Our integrated model presented below will follow Figure 2, but will integrate the open innovation concepts of Chesbrough into each of the stages as appropriate. The use of this innovation model by LSPs will result in a structured and disciplined approach that integrates open innovation concepts as a part of the process whenever they are trying to develop services to offer to a significant market segment. This is in contrast to the evolutionary process that Hertz and Alfredsson (2003) discovered in their research. At first glance, the process appears to be both long and complex. However, practice in using it should result in LSPs being able to implement it fast enough to respond to their market places.

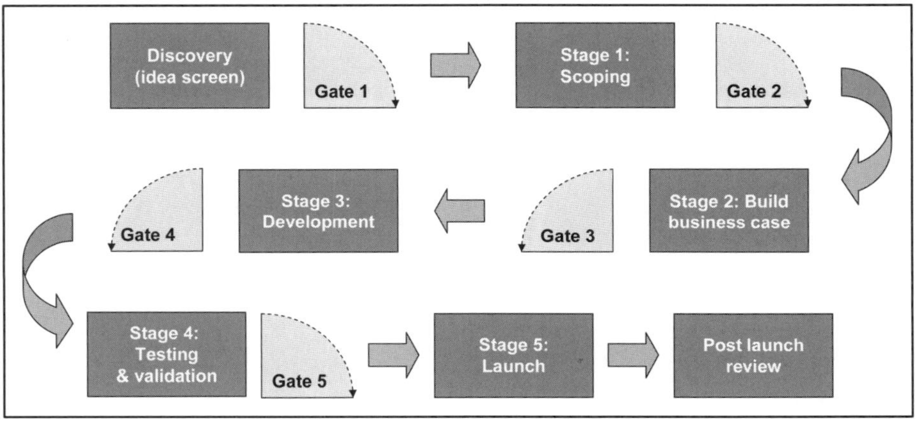

Figure 2: A Sample Stage-Gate™ Process

One basic idea of the Stage-Gate™ process is that the criteria for successfully passing through the early gates are much easier than for the later gates. This is to ensure that good project ideas are not rejected prematurely because little information is yet available to evaluate them. It is also true that projects do not consume large amounts of resources during the early phases, so it is reasonable to pass many of them through the gates for further evaluation. As later gates are reached, a decision to pass the project through to the next phase means a decision to invest more resources into the project, and so the criteria for evaluation become much more rigorous and demanding. This ensures that only the best projects should pass these more difficult later gates, and the weaker projects are killed. In this way organizational resources can be focused on the potentially best projects to try to rapidly bring with the greatest potential to come them to fruition.

Each gate in Figure 2 has a defined structure to include:

- A set of required deliverables defined at the previous gate
- Criteria against which the project will be evaluated to include both "must meet" and "should meet" criteria
- Defined outputs including a decision to go/kill/hold/recycle/outsource this project; if a go, an approved action plan and deliverables for the next gate.

Discovery

The discovery phase prior to Gate 1 is critical; without many great product ideas, no great products are likely to be developed. Many companies formalize this process and call it *ideation*. A host of internal and external sources should be used to discover as many good new product ideas as possible. Cooper (2001) provides many examples of potential sources for ideas:

- Use the own company's product innovation and technology strategy to help focus the efforts
- Use both bottom-up and top-down approaches
- Develop alternate scenarios of the future including best case and worst case scenarios
- Try to utilize the creative talents of the entire organization
- Assign one person to serve as a focal point to stimulate all the idea sources
- Set up an "idea bank" to hold the results of the brainstorming activities until they can be evaluated, so they don't get lost
- Immerse a cross-functional team in an area for a period of time with the explicit objective of generating new product/service ideas
- Publicize the ideas across all business units so that many perspectives may contribute to the idea list
- Provide time for people to engage in creative thinking for new product/service ideas
- *Determine if there are disruptions in the customer's industry that provide ideas*
- *Utilize the voice of the customer for new product/service ideas*
- *Partner with leading or innovative customers*
- *Identify fundamental research breakthroughs that might change the competitive landscape*

- *Look at what the competition is doing*
- *Go to trade shows and collect information aggressively*
- *Review trade publications from around the world*
- *Review patents that have been filed*
- *Work with suppliers to mine their ideas*
- *Find out what universities are doing in their research programs.*

The italicized entries above go beyond what Chesbrough (2003) calls "closed innovation." Chesbrough (2003) would add government sources and professional societies to the list. He would also stress that relationships should be built with many of these external entities to "identify and connect" with them for ongoing interactions, so there is a constant "buzz" of activities in innovation.

Gate 1 in Figure 2 is intended to be a "gentle screen" of the ideas. According to Cooper (2001), the criteria used for this gate could include strategic fit with the company, feasibility, size and attractiveness of the opportunity, product differentiation and advantage, fit with the company's existing resource base, and fit with company policies.

Scoping

The objective for Stage 1: scoping is to spend only a small amount of money to gather additional information to allow for a re-examination of the idea at Gate 2. In Stage 1, preliminary estimates of the market, technical, and financial possibilities for the idea are prepared. The market analysis at this stage is not rigorous market research. Instead, sources of information are used that are easy to access and inexpensive. Cooper (2001) lists the following as examples of information sources for this phase:

- Internet searches
- Company library
- Internal reports
- Key customers
- Focus groups
- Advertisements

- Fellow employees in company
- Consulting and research firms
- Financial institutions
- Government agencies
- Industry experts
- Editors of trade magazines
- Trade associations

According to Cooper, the company's technical staff including R & D, engineering, and operations would conduct the preliminary technical analysis in Stage 1. However, an open innovation model would suggest that the possibility of incorporating supply chain partners, even including competitors if appropriate, at this stage of the process and to have assignments for various entities to complete as part of the technical analysis. Cooper (2001) asks the following key questions to assess the technical viability of the project at this early stage that should be answered regardless of the parties involved:

- Approximately what will the product/service requirements or specifications be? (Note that the operating definition may still be vague and fluid at this early stage)
- How would these requirements be achieved technically? Is there a foreseeable technical solution? Is invention or new science required?
- What are the odds that the product/service is technically feasible? At what cost and time?
- Is the technological capability available to develop it? Or is it necessary to seek a partner or outside supplier for some development work?
- Can the product/service be developed and offered? How, with what equipment, and at what cost? Or should outsourcing or a partner be considered?
- What intellectual property and product/service regulatory issues are involved? Note that regulatory and patent issues are introduced and considered at this early stage
- What are the key technical risks? And how might each be handled?

The last component of Stage 1 is the preliminary business and financial assessment. The business assessment focuses on whether the company has the core competencies to undertake the project in its entirety, or whether a partner is needed to outsource some of the work, or outsource the entire project. Estimates of sales, costs, and investment required are made, but with the realization that they are only guesses at this time. Based on the combination of the three assessments, a recommendation and a project plan are developed for Gate 2. This project plan should include recommendations of an open innovation nature, including the roles of various value chain partners.

User Needs-and-Wants study

Cooper considers Stage 2 as the most critical to the success of the project, and the stage that is most often done poorly by companies. There are a number of activities that need to be done to thoroughly explore the business case for the idea. One of the first is the user needs-and-wants study. This study may start out by using less in-depth procedures such as focus groups, but should include a significant, systematic survey of a large number of potential customers with in-depth analysis at a later time. Techniques such as quality function deployment (QFD) may be used in this stage. The focus needs to be on the benefits that the product/service will provide to the user as its value proposition, that then lead to a definition of the attributes and features of the product/service that will provide the performance. A significant amount of face-to-face interaction with the customers is imperative in this stage. The objective is to live virtually in the customers' shoes so as to observe the needs and wants that they have that could be met by the product/service. This is consistent with the open innovation concept.

Competitive Analysis

A second area for in-depth review at Stage 2 is the competitive analysis of both direct and indirect sources of competition. Indirect sources may be identified by viewing the industry in broader terms. For example, a newspaper company can obviously view its competition as other newspaper companies. Taking a broader view, companies that sell magazines are also competitors, since they too distribute information. Television, radio stations and internet sources also distribute news and other information, so it would be short-sighted on the part of a newspaper to view only other newspapers as the competition. The analysis of the competitors should be as thorough as the analysis of the needs and wants of the customers. The LSP should (1) understand the competitors' products and their strengths and weaknesses, (2) understand the value proposition that each competitor presents to the customer, and (3) understand the organizational strengths and weaknesses of the competitors so that their response to the market offerings can be anticipated. In other words, the LSP should develop a strategy that maximizes the strength of the offering compared to all others, and that creates a source of sustainable advantage for at least an intermediate period of time. This view of competitive analysis is consistent with open innovation concepts.

Market Analysis

A third area for exploration in Stage 2 is the market analysis. The user needs-and-wants study and the competitive analysis provide important information to support the market analysis. All the information sources listed in the scoping category add input for the market analysis, but more is needed. The size of the market, current trends, and future expectations about growth or contraction are likewise needed. A market segmentation analysis also needs to be done along the same lines as the total market analysis. We need to understand who the buyers are in each segment and obtain detailed information on their buying habits and processes so we can develop ways to reach them. This market analysis is clearly a part of the business model functions in Chesbrough's open innovation model.

Technical Investigation

The next area for consideration is a detailed technical investigation. With its preliminary product/service concepts in mind, the company has determined whether it is both technically and economically feasible to design, development, and offer the product/service. The central internal players in this analysis may include marketing, R & D, engineering, operations, and legal. External value chain partners may similarly contribute critical parts of this analysis. Quality function deployment can be a very valuable tool at this stage to organize and present the primary information of both the market or customer specifications and the technical requirements to all parties involved.

Concept Testing

The next step in the process is concept testing with the customer. The presentation needs to be neutral and objective, not a sales pitch, so as to get an accurate gauge of customer enthusiasm. Focus groups are recommended in the early part of this activity, followed by one-on-one presentations to a representative sample of customers. If the product/service is to be test marketed, the LSP should try to observe the test, including the customer's participation. Thorough documentation of the customer interaction and feedback to facilitate later product/service evaluation should be done. Supply chain partners may participate in this process, but the information to be sought as a result of the concept testing includes:

- An assessment of the strength of the customer's interest in the product/service and why the customer has this interest
- An understanding of the customer likes and dislikes about the product/service
- A comparison from the customer of the proposed product/service versus the one the customer is now using, with explanations of why they prefer certain attributes of one or the other
- How much the customer might be willing to pay for the product/service
- How likely the customer is to purchase the product/service for a specific price(s)
- Information to help position the product/service in the marketplace.

Business and Financial Analysis

The next activity in this phase is the business and financial analysis. With the information gained from the earlier activities in this stage, the company is now able to produce useful estimates of revenue versus costs, including investment and operating costs, for the product/service. Standard financial analysis should be performed using these estimates including pro forma estimates and discounted cash flow computations. Cooper makes a strong argument that product/service development decisions must use a portfolio approach. A portfolio approach takes a holistic view of a set of investments with the objective of maximizing the entire portfolio value. In an open innovation setting, this analysis has to include how each value chain partner is to participate and benefit from the product/service offering.

Plan of Action

The final activity in this stage is to prepare a plan of action for the next stage based on everything learned so far. The plan of action includes a recommendation as to how to proceed (go/kill/hold/recycle/outsource), a detailed development plan (described later), a preliminary plan for operations and the supply chain as needed, a preliminary marketing plan, and a first plan for Stage 4, testing and validation (also described later).

Development

The next stage is development that results in a concrete product/service. This stage may be an iterative one since the parties may have learned that there are some needed adjustments to the product/service design, or the external situation could have changed so as to require some mid-course corrections. It is important to maintain close contact with the customers throughout this stage to ensure that the resulting product/service really meets their needs. The important thing is to involve the customer in deep and meaningful ways during this most critical period.

Since development is often the longest part of the new product/service process, it is imperative to manage development times so as to keep them as short as possible, consistent with a quality process. Cooper (2001) provides many suggestions for shortening the development time. Superb project planning and execution will be critical in an open innovation network of partners to ensure a timely completion of the project, with the activities completed at a high level of excellence.

While the product/service development is proceeding, other planning aspects of the project need to be carried out. These include updating the market review and development and specifying the detailed market launch plan, dealing with intellectual property and product/service regulatory issues, completing the detailed design for the operations and supply chain activities, updating the business and financial analysis data, and developing the standard outputs to guide Stage 4. Again, various partners may carry out assignments to complete these activities.

Testing and Validation

The Stage 4 testing and validation process is a customer-focused process. The customer should be involved while the development is proceeding, rather than waiting until the product/service is finished. The objective of the testing and validation process is the ultimate validation of the entire project including the product/service itself, and its operations and marketing plan. This process needs to be carefully documented since it is the basis for a decision on whether to launch the product/service at the succeeding gate. Cooper (2001) provides many additional suggestions for the testing and validation phase. An important activity in an open innovation network at this stage is to define the position of each partner within the value network to facilitate each partner's decision concerning continued participation at the launch stage to follow.

Launch

The final stage in Figure 2 is the launch. Cooper discusses this stage on the marketing issues without considering the production and supply chain issues. The marketing plan for the launch should include the marketing objectives, strategies, and programs. Even though this is the last stage in the process, work has been going on since early in the project to support it, and has been refined as the project develops. The marketing objectives should establish criteria for making decisions, be operational so they can be measured, and have a timeline associated with them. Sales volume and profitability objectives would typically be included. Even though these goals are now based on an updated market analysis that includes the testing stage, there may still be considerable uncertainty about them, and some flexibility needed by the company in moving forward. A fresh look needs to be taken at what Cooper calls "macroenvironmental analysis." This involves an investigation of the following elements wherever the product/service is to be marketed so as to estimate any impact these variables may have:

- General economic situation
- Governmental situation
- Demographics
- Social developments and trends
- Technological developments

Similarly, an assessment of the company's strengths and weaknesses is prudent prior to the final assault on the market. All supply chain partners will need to perform this assessment as concerns their continued participation. The results of these assessments may cause some changes to the way the offering is marketed. With this information in hand, the target markets for the launch, and the segments within those markets may be selected. These are based on standard marketing considerations that should have been included in the earlier versions of the marketing plan, but now may be updated on the basis of the latest and best information.

Post-launch Review

The product/service strategy should be finalized at this stage based on the market segments and product differentiation decisions that have been made. The value proposition to the customer for the product/service should be clearly and accurately stated. Both the communication and advertising plan and the channels for getting both the product/service information and the product/service to the customers should be finalized, including any use of the internet and e-marketing. Enough detailed information should be available to establish the pricing strategy and its details based on the preliminary market acceptance and the competitive situation. All marketing plans should be consistent with corporate strategies of the partners. The sales forces should be active participants in this stage of the process, both to be trained about the new product/service and to provide input and suggestions about all aspects of the marketing plan. They must be fully on board prior to the launch. With updated information about the marketing plan including its costs, a final update to all the financial analysis should be completed to provide a strong basis for the final go/kill decision. Figure 2 concludes with a post-launch review. The purpose of this review is to keep learning about the innovation process in order to continually improve it.

This integrated innovation process should provide a central weapon for an LSP to use to develop innovative services to enhance its value to its customers. But we have not yet provided the focus for the use of this important weapon to help the LSP achieve essential link status. In the following section, we introduce the concept of operating models for market leaders to fill in the picture.

5.4 Three Models for Market Leaders

Over a three-year period, Treacy and Wiersema (1995) studied 80 companies serving more than three dozen markets at different levels of excellence in order to understand why some succeeded while others did not. They noted three distinct operating models that market leaders followed, and that market leaders chose one of the three as their basis for differentiation, and provided threshold performance for each of the other two. Focusing on the chosen model resulted in the market leadership position attained. The three models included:

- Operational excellence
- Product leadership
- Customer intimacy

LSPs can use these models to decide how they want to be positioned in the markets that they serve, and to help them tailor their innovation efforts to their strategic choices. The resulting innovation efforts should help them develop the competencies they need to achieve the status of the essential link within their markets.

Operational Excellence

An operationally excellent company provides reliable products/services at competitive prices, delivered with minimal difficulty or inconvenience. They offer a quality, price, and ease of purchase package that no one else can match. Their focus is on low prices, and standardized assets and efficient operating procedures are the backbone of the operationally excellent company. Variety kills efficiency; so the product line is minimized. The emphasis in people management is the team; free spirits are not welcome. They strive to minimize overhead and continually reengineer business and operating processes. They are high tech at the core, centralized, regimented, and finely automated. They are passionate about measurement and cost control, and streamline customer service. They focus on growth and large volume market segments. Because their product variety is limited, they do not please every customer. Their management systems are focused on high speed, integrated transactions that conform to the norm. Their culture abhors waste and rewards efficiency. Fast food restaurants are an example of companies with a primary focus on operational excellence. For LSPs, a focus on flawless execution of basic warehousing and transportation services so as to be the low cost provider would illustrate an operational excellence focus.

Product Leadership

A company that focuses on product leadership provides products that continually set new standards for state of the art. Their obsession is to create exciting new products that provide greater value to their customers than their competition can. They require a fluid organization that attracts talented and curious people. They have to excel at quick commercialization of their ideas. They educate their customers to accept new products. They use a portfolio approach to try to place bets on winners. They have flexible organizational structures that allow them to manage eccentrics. Their customers pay premium prices for the unique products. They engage in aggressive marketing, selling, and price management over the product life cycle and try to find and exploit all market niches. They focus on the core processes of invention, product development, and market exploitation, and measure themselves using metrics such as the per cent of sales coming from new products. Sony, 3M, and Apple are examples of firms that focus on product leadership as the key operating model. For LSPs, we interpret product leadership in two distinct ways. First, the LSP may provide a truly unique service that differentiates it from most of the marketplace in significant ways. An LSP that has truly integrated information systems across a manufacturing customer's ERP system, transportation management systems, and warehouse management systems so as to facilitate processing customer orders, might illustrate this absolute product leadership approach. A second interpretation is more subtle, but we believe much more prevalent. It is based on the simple concept of market segmentation. An LSP might observe that a particular industry has a certain "package" of needs, no one of which is unique. However, putting the package together to cater to the particular industry and thus to differentiate the offering in the marketplace as a "one stop shop" *for this particular industry* could also constitute a product leadership position, albeit on a smaller scale. A lot of niche providers would fall into this category. The LSP industry would call this "verticalization." Our approach would cause this to happen as a result of careful design, however, and not as a more evolutionary, accidental process.

Customer Intimacy

A customer intimate company sells its customers a total solution, not just a product or service. Success in this case is a cooperative success between the two partners. Such companies do not necessarily offer the lowest price or the best products. They attend to more of their customers' needs to provide greater value. They may personalize service, customize products, or both. They may take responsibility for their clients' successes, and may help revamp client business processes that involve the use of their products or services. They become experts at their customers' businesses. They are client-driven, and measure their success by their clients' successes, and by their company's share of the client spending. They have a decentralized organization for greater responsiveness. They may organizationally "merge" their employees into the customer's site. Since their customers acquire the expertise that the company develops, they constantly search the environment to continue to add new value to the customers' operations. They are obsessed with solution development, results management, and relationship management. Their management systems are dedicated to producing results for carefully selected and nurtured clients. Their culture thrives on deep and lasting client relationships. Consulting firms are quite often dedicated to this discipline of customer intimacy. In their book, Tracey and Wiersema (1995) use Airborne Express as an example of a customer intimate company, in contrast to companies like FedEx and UPS. The question arises as to how to differentiate a customer intimate LSP from the subtle form of product leader mentioned above. We believe the useful distinction is that the subtle product leader will not go outside the range of capabilities that they have chosen to define themselves in the marketplace, whereas a customer intimate firm will find and obtain any service for which its client has a need; so the difference is one of scope.

5.5 Market Leader Models and LSPs

The three models for market leaders introduced in the preceding section shall now be applied to the LSP industry.

Operational Excellence

Yeung (2006, p. 121) found that "*timeliness of the service, pricing, and the quality of delivery of 3PL service providers are positively related to the user's logistics and/or export performance.*" These are clearly measures of operational excellence. Ramaswami (2006, p. 33) cites survey results relating to standardization: "*Users expect 3PLs to leverage their solutions and technology across several customers and offer deep industry knowledge at a reasonable cost.*" Quinn (2006) mentions that there is still downward pressure on prices, and that shippers expect a service package including pickup, cross docking, shipment, breakdown, and delivery all for one price; again, clear examples of operational efficiency. In Cork (2006, p. 16), John Hawksworth, commercial director of the Widdowson Group, is quoted as saying: "*We will still retain and grow our transport offering, because that's always been a specialist part of our business. And, to be honest, if you're good at transport, you can still make the right margin from it.*" Atkinson (2007) lists how LSPs can add value to include making the best use of space, minimizing time, distance traveled, errors, and inventory needed to provide high levels of fulfillment. In an article on how to source logistics services, Gould (2003) describes one type of purchases as "leveraged." This type of purchase is made to minimize purchase price, and consists of grouping items for purchase from a supplier in order to gain additional discount. Lieb (2005, p. 22) reports survey results that even though service offerings are broadening, "transportation and warehousing revenues still dominate in most 3PL companies." Eyefortransport (2007a) lists 14 reasons why customers did not renew the contracts of their 3PL, and the percentage of respondents who cited each reason. The top seven reasons were listed by a range of 34 to 61% of the respondents, whereas the bottom seven reasons were listed by a range of only 3% to 18%. Five of the top seven reasons were operational excellence criteria. It is certainly not surprising that it is easy to find ample evidence to support a model of operational excellence for LSPs. In a network of collaborative and complementary LSPs, there is still a role for an LSP to focus on a narrow range of core competencies and to be successful in doing that. However, the LSP should be forewarned that competition in such markets is characterized as commoditized services, and there is likely to be never-ending downward pressure on prices.

Product Leadership

Fulconis et al. (2007, p. 114) describe a concept for a 4PL that "*requires here a centralized IOS (Inter-organizational Information System) that provides control over all potential activities related to the physical acquisition, storage and transfer of goods.*" Such an integrated information system would create a product leadership position for a 4PL. Power et al. (2007, p. 228) explain the practical implications of their study of outsourcing to 3PLs: "*The practical implication of the study is that it provides support for the potential for an effective set of arrangements with 3PLs to help organizations to move out of the world of pure cost based competition, and into one where they could compete on multiple dimensions.*" This illustrates the subtle form of product leadership based on market segmentation. In their study of LSPs in Internet Supply Chains, Rabinovich and Knemeyer (2006) characterize the roles that LSPs can play for their clients, such as providing "hub functionality" to make the sellers more responsive to changing market requirements. This involves performing multiple roles for the seller; more than half of the sellers in their study had LSPs performing three or more functions. This functionality can bring a service differentiation for the seller vis-à-vis its competitors that can generate better margins. Again, this hub functionality illustrates the subtle form of product leadership. They do point out, however, that the use of LSPs for services requiring high asset specificity is unlikely to favor the sellers.

Wu (2006) assessed technological innovations in patenting for 3PLs. He concluded that patent statistics can provide a good picture of the state of logistics innovation, and that LSPs can innovate by making good use of the patent data. Such leading edge innovation should result in the more unique form of product leadership, rather than the more subtle form based on market segmentation. Quinn (2006, p. 6T) reports on a survey of merger and acquisition activity in 3PLs. He quotes Evan Armstrong of Armstrong & Associates, "*Companies that acquire to increase specific capability levels seem to be more successful in their results than those that are simply trying to increase network scope.*"

Such firms appear to be assembling the right mix of capabilities to practice the subtle form of product leadership via market segmentation. Cork (2006, p.17) reports on a variety of service offerings of logistics providers. She quotes Jeff Anderson of Wincanton, "*At Wincanton, we've found that by incorporating packing, reworking and inspection activities into traditional warehousing services and by investing in dedicated systems and facilities, demonstrable added value has been achieved in our customers' supply chains.*" This quote spells out the essential mix that Wincanton chose to differentiate itself as a product leader.

Atkinson (2007) describes mass customization for Xerox and a "clean room" for assembling and filling of food and non-food consumer products as illustrations of service innovations geared toward differentiation, and thus the subtle from of product leadership. He explains that the ideas for such innovations come from both the customer and the engineering departments of the 3PL firms. Returning to the Gould (2003) article on outsourcing logistics services, a second category of purchases is "critical purchases." These purchases are used to reduce risk, are critical to the ongoing operations of the company, and are in limited supply. An LSP who can deliver consistently in such an environment becomes a product leader. Lieb (2005), in reporting the results of a 3PL industry survey, lists several potential areas for a 3PL to establish itself as a product leader. One of these areas is globalization, which has been an area of ongoing concern for international shippers. An interesting one was customer selectivity (i.e. focusing on companies of certain sizes in certain industries, with the result being higher margins and yields). This was an important point for emphasis by Tracey and Wiersema (1995); by choosing the market discipline (operating model), a company is choosing (or rejecting) its partners.

Still another area was the integration of technology, including the vexing question of information technology. Once again, there is ample evidence that both the direct and the subtle forms of product leadership can be easily identified as a part of the 3PL landscape. To succeed at the subtle form, an LSP must choose the right mix of capabilities that match a significant market segment, and deliver the mix at a high level of excellence. To succeed in the more absolute form of product leadership, such as globalization, the LSP must overcome significant uncertainty to solve the problem and it hereby establishes itself as the market leader. Capabilities of such market leaders are quickly copied and become a part of the standard product offerings of many competitors. Staying ahead of the game is therefore a major challenge for product leader companies.

Customer Intimacy

We also found ample evidence of customer intimacy in the LSP arena. Fulconis et al. (2007, p. 114-115) describe a 4PL as "*a client-driven solution that has tremendous potential in the European market.*" Panayides (2007, p. 78) used this research to conclude, "*The results support the importance of relational capabilities, and indicate the value of organizational learning in terms of improving logistics service effectiveness and consequently firm performance.*" Rabinovich and Knemeyer's (2006, p. 106) study of LSPs in Internet Supply Chains concluded: "*It makes sense to set up a link with an LSP if doing so would expand the margins an Internet seller obtains – increased margins come from gaining access to profitable and loyal customers by leveraging the LSP's relationship networks.*" In such situations, the criterion that the LSP is taking some responsibility for the success of its customer (the seller) is clearly met. In Yeung's (2006, p. 121) study of the impact of 3PLs on the performance of their customers, he admonishes 3PLs "... *to downplay offerings of customized supplementary ser-*

vices unless they are engaged in long-term relationships with the users." Such long-term relationships typically signify a customer intimate relationship.

Artman and Martha (1997) characterize the development of long-term relationships between shippers and 3PLs. They point out that these relationships require years of effort to allow sufficient trust to develop between the partners. Cork (2006, p. 16) provides two examples of customer-intimate relationships. The first was from commercial director John Hawksworth of Widdowson Group. He described his company's foray into contract packing: "*We dedicated a 70,000 square foot warehouse on our main Leicester facility to do this, so it is now effectively a processing shop. The more you can offer your customers, the more you tie them in to your business.*" A second example of a customer intimate relationship was between Lucite International and its LSP, Bibby Distribution. Lucite invited Bibby to sit in on the project development work for its new IT system. Gould (2003) defined a "strategic" purpose, characterized as one that creates competitive advantage, and requires a high level of trust between the LSP and the customer. Lieb's (2005) 3PL survey noted that two popular trends were customers desiring to outsource more and more of their logistics services, and an increasing desire for "one-stop shopping." Both of these developments suggest customer intimate relationships with the LSP.

In the Eyefortransport (2007a) survey, one of the top seven reasons listed for non-renewal of 3PL contracts was relationship management issues. This could certainly signal a failing customer intimate relationship. In a separate Eyefortransport (2007b) survey of outsourcing logistics, 77% of respondents indicated that they considered their 3PLs as supply chain partners rather than as mere service providers. Such partnering arrangements often lead to longer term contracts and associations that allow customer intimate relationships to flourish. An LSP that focuses on customer intimate relationships must find and develop customers that are amenable to such an approach; that is no easy task. The LSP can never rest on its laurels, since there will be a natural transfer of its expertise to the client because of their close interaction. The customer intimate LSP must foster a highly flexible and adaptive organization, and be willing to learn about new areas very quickly as the need arises.

5.6 Summary

In this paper, we have presented an integrated innovation model that is structured, disciplined, and open. We summarized three operating models for market leaders: operational excellence, product leadership, and customer intimacy. We drew from recent publications to illustrate the relevance of the market leader models to the LSP industry. We argue that the integrated innovation model used in conjunction with the market leader concepts have not been recognized and exploited by LSPs, and that doing so could help LSPs become the essential link in their markets. If the integrated innovation model is likened to an automobile, then the market leader models gives it directions about where to go. If the operating model is for operational excellence, then the innovation process should focus on implementing the technology and business processes that lead to the most efficient operations. If the operating model is product leadership, then the innovation efforts should pioneer original innovations and/or define and hone just the right mix of capabilities to match a particular market segment of customers. Finally, a customer intimate market discipline means that there should be a focus on relationship management, customer retention, quick learning and adaptability, and opportunity identification and exploitation. In other words, the market leader model provides the focus for innovation efforts, and the clarity of that maximizes the chances for becoming the essential link. Figure 3 provides some examples of LSP services that fit into each of the three operating models. Of course, many more could be listed. The figure also lists the key dynamic for each of the three operating models to help leaders of LSPs understand the processes on which they will need to concentrate.

Market leadership model	Examples of LSP service offerings	Key dynamic
Operational excellence	Pickup and delivery Cross docking Fulfillment Basic warehousing	Flawless execution
Product leadership	"4PL" Clean room assembly Integrated information Critical purchases Globalization support	Original innovation
Customer intimacy	Access to LSP customers Joint IT development Home delivery "End-to-end" solutions Supply chain consulting	Trust and flexibility

Figure 3: Matrix of Market Leadership Examples and Key Dynamics

We need to make clear, however, that for a large LSP, different parts of the organization may follow different market disciplines. The disciplines should apply at a strategic business unit level, however, and provide the focused innovation we have described. We have also emphasized that sustaining the essential link status in the markets served requires a continuous commitment to focused innovation and improvement.

References

Artman, L.; Martha, J. (1997). Establishing Long Term Relationships, *Transportation and Distribution*, 38 (3), 69-72.

Atkinson, W. (2007). Value Proposition, *Material Handling Management*, 62 (5), 20-26.

Chesbrough, H.W. (2003). *Open Innovation: The New Imperative for Creating and Profiting from Technology*, Cambridge, MA: Harvard Business School Press.

Cooper, R.G. (2001). *Winning at New Products: Accelerating the Process from Idea to Launch*, 3rd Edition, New York: Basic Books.

Cork, L. (2006). Not just Flavour of the Month, *Works Management*, 59 (12), 16-18.

Eyefortransport (2007a). *The North American 3PL Market – A brief analysis of eyefortransport's recent survey*. Available: http://www.eyefortransport.com/3pl2007/rep07.pdf.

Eyefortransport (2007b). *Outsourcing Logistics USA 2007 – Best practices for managing 3PLs*. Available: http://www.eyefortransport.com/3PL.

Fulconis, F.; Saglietto, L.; Pache, G. (2007). Strategy Dynamics in the Logistics Industry: A Transactional Center Perspective, *Management Decision*, 45 (1), 104-117.

Gould, S.A. (2003). How to Source Logistics Services Strategically, *Supply Chain Management Review*, 7 (5), 48-54.

Hertz, S.; Alfredsson, M. (2003). Strategic Development of Third Party Logistics Providers, *Industrial Marketing Management*, 32 (2), 139-149.

Lieb, R.C. (2005). The 3PL Industry: Where it's Been, Where it's Going, *Supply Chain Management Review*, 9 (6), 20-27.

Panayides, P. (2007). The Impact of Organizational Learning on Relationship Orientation, Logistics Service Effectiveness and Performance, *Industrial Marketing Management*, 36 (2007), 68-80.

Power, D.; Sharafali, M.; Bhakoo, V. (2007). Adding Value Through Outsourcing: Contribution of 3PL Services to Customer Performance, *Management Research News*, 30 (3), 228-235.

Quinn, J.P. (2006). 3PLs Hit Their Stride, *Logistics Management*, 45 (7), 3T-11T.

Rabinovich, E.; Knemeyer, A.M. (2006). Logistics Service Providers in Internet Supply Chains, *California Management Review*, 48 (4), 84-107.

Ramaswami, R. (2006). Spotlight: 3PLs, *Multichannel Merchant*, January 2006, 33-33.

Treacy, M.; Wiersema, F. (1995). *The Discipline of Market Leaders*, Reading, MA: Addison-Wesley Publishing Company.

Wu, Y.J. (2006). Assessment of Technological Innovations in Patenting for 3rd Party Logistics Providers, *Journal of Enterprise Information Management*, 19 (5), 504-524.

Yeung, A.C. (2006). The Impact of Third-Party Logistics Performance on the Logistics and Export Performance of Users: An Empirical Study, *Maritime Economics and Logistics*, 2006 (8), 121-139.

Chapter 6

An Audit Tool for Innovation Processes of Logistics Service Providers

Christian Busse and Stephan M. Wagner

Abstract

Even though the necessity for industry-specific considerations has been recognized, research on innovation management at logistics service providers (LSPs) is only beginning. Academics and practitioners lack audit tools allowing for the analysis of innovation processes at LSPs. By means of conceptual work and the use of action research, such an audit tool was developed and is presented here. It consists of four phases with twelve activities and is accompanied by considerations on centralized or decentralized allocation of actors, on IT and management tools and on success factors. All of these deliberations occur for each of the activities. This audit tool is an instrument for further research, and the management at LSPs is given a tool for the development of improvement measures.

6 An Audit Tool for Innovation Processes of Logistics Service Providers

6.1 Introduction

Innovation activities and innovation output at LSPs are still rudimentary; innovations are frequently not actively managed (Wagner, 2007). Apparently, the concepts developed within the field of innovation research do not suit the industry or they cannot be implemented effectively by the LSPs. Thus, an industry-specific perspective has to be applied. Due to the vast possibilities of research in this field and the large research deficit, it is vital to narrow the subject to a single focus area. To this aim, we follow van de Ven (1986, p. 591) who states: "From a managerial viewpoint, to understand the process of innovation is to understand the factors that facilitate and inhibit the development of innovations." Hence, this article is primarily concerned with the analysis of processes of innovation development at LSPs and investigates the following research question: *"Which model of innovation development should be used by logistics service providers (LSPs) as an audit tool, i.e., to accomplish analytical tasks?"* Accordingly, the model presented here is not to be used as a tool which can be rolled out through an LSP's organization in its entirety (it is too complex for that), but to allow for a structured analysis of innovation development. Such an analysis might have to be carried out by, for example, an innovation manager looking for bottlenecks or by a management consultant reviewing a long-term business plan. The audit tool in itself is supposed to contribute to the success of innovation management at LSPs.

An action research approach was chosen as research method. This approach allows not only the description of the existent processes; it also concurrently helps to generate a normative model.

6.2 Methodology

Action Research

Given the exploratory nature of our research and the complexity of the individual research subjects, it was reasonable to use a flexible methodology on a small scale. We chose the action research approach which is increasingly used in logistics research (e.g., Bichou and Gray, 2005; Koplin, 2005; Middel et al., 2005; Müller, 2005; Prockl, 2005).

According to Müller (2005), action research can be identified by the following attributes: relation to practical problems, action, discourse orientation, embeddedness in practice, function of the researcher as an agent of change, and relation to dialectics. The aim of an action researcher is to obtain a holistic understanding of the situation, not only of a few variables. Through close interaction with managers in charge it is possible to arrive at a deep situational understanding which is especially helpful for establishing theories (Westbrook, 1995). In action research it is vital to document the proceeding in detail. The procedure is characterized by iterations, heuristics and feedback. Action research requires a proactive management of the research process itself. A convenient instruction for research design can be found in Westbrook (1995). Figure 1 shows the advice and its use in this research project.

This agenda led to detailed and, according to the experts involved, valuable findings, reaching far beyond the existing standards in this research area. Criticism could arise from the fact that the researchers went beyond their usual observatory role by becoming involved in action research. Specifically, they proposed topics, moderated, discussed and thus affected the model itself. However, it is precisely this approach which allows not only the development of a descriptive model but also one which claims to contribute to an improvement of innovation processes management.

Advice	Coverage
Agree on "the problem areas [...], but do not pre-specify the solution"	▪ Open discussion about possible phases, gates, activities, responsibilities for assignments, tools and pitfalls
"Seek multiple viewpoints […], no one [..] will see the whole picture"	▪ Group discussion among six to eight experts ▪ As far as possible, two experts of each partner corporation
"Record in simple semi-standard formats"	▪ Presentation of figures in Microsoft PowerPoint ▪ Completion of figures and tables in Microsoft Excel by the experts
"Have the company check your write-ups"	▪ Use of mailing list ▪ Discussion of the figures and graphs at the workshops
"Prefer data to opinion"	▪ Realization impossible, because of the competitive environment amongst the experts' companies
"Remember that opinions are also data"	▪ Documentation of process and discussion with detailed protocols
"Use an assistant to do much writing-up"	▪ Documentation of discussion by currently inactive person
"Give some thought to the right frequency of site visits"	▪ Agreement about the frequency and overall length, upfront

Figure 1: Action Research Design and Implementation (Source: Westbrook, 1995)

Expert Panel

An expert panel, consisting of executives from the logistics service industry was selected to work on the topic. The conception occurred at the beginning of 2006. The intention was the development of recommendations on how to manage the innovation process of an LSP in practice. Hence, the companies as a whole were subject to examination.

In order to accomplish an industry-specific examination of the topic, it was useful to limit the research to LSPs. Ideally, every company should be represented by two participants in order to reveal possibly differing perspectives. Seven executive managers of four large, internationally-operating LSPs were permanent members of the expert panel, which met in the second half of 2006. The round met four days in total. The workshops were prepared in advance and summed up in detail afterwards.

The participants belonged mainly to the second management level of a German-speaking organization or its corporate headquarters. A broad area was covered, reaching from product development and management functions to corporate development, key account management, and to chief of operations. The research team consisted of two people, one moderator and a second researcher acting as a co-moderator and devil's advocate during the discussions.

The decision on the topics of the workshops was dynamic. During each workshop, the experts decided on the topics of the following one. The key criterion for the choice on a subject was its potential for practical implementation of the results. Out of this requirement, the workshops focused on the processes of innovation creation at LSPs and on the development of a convenient model.

Requirements for an Audit Tool

Ten requirements for the audit tool were established. The first three were derived from General Model Theory (Kühne, 2005), based on the central features of any model. First, the mapping feature: Models are maps of certain originals which may have an abstract character themselves. Second, the reduction feature: Only relevant properties of the original are mapped into the model. Third, the pragmatic feature: Models fulfill a surrogate function for their users. A model's users determine its purpose.

Further requirements are a process-related connectivity with a company's innovation management system and the necessity to include procedural learning and improvement. Both requirements can be established by means of a comparative analysis of established process models (e.g., Bowers, 1989; Cooper and Kleinschmidt, 1993).

Finally, features which most LSPs possess determine the design of an audit tool for them. These are their often down-to-earth character, their usually comparatively low level of education, their often decentralized organizational structure, their customer specificity and finally, the service character of their logistical offerings (Chapter 1).

Each of the requirements established here will be considered individually in the following chapter.

6.3 Results

For the expert panel, it was necessary to embed innovation processes into a comprehensive innovation management system. Such a framework helps to understand the surrounding conditions for the sequence of events in innovation processes. There are some frameworks in the academic literature, but most of these contain elements which at least intuitively are difficult to link to a procedural understanding (e.g., Adams et al., 2006). Hence, within the expert panel discussions, a proprietary framework was set up which includes a number of process-related innovation management elements that go beyond the management of individual innovation processes. It integrates the modeling of the individual innovation processes.

We begin with the description of the innovation management framework. Afterwards, we will look at individual innovation processes, their different phases and their specific design for LSPs.

Innovation Management Framework

Figure 2 depicts the innovation management framework which forms the basis of our analysis. Innovation management of LSPs can be implemented and established within this framework. Modifications can be made in an incremental way without the need for structural breaks.

The modification of the elements of the framework is ensured through continuous learning which is enabled by the "cumulated innovation process experience" element that collects the experience gained over time.

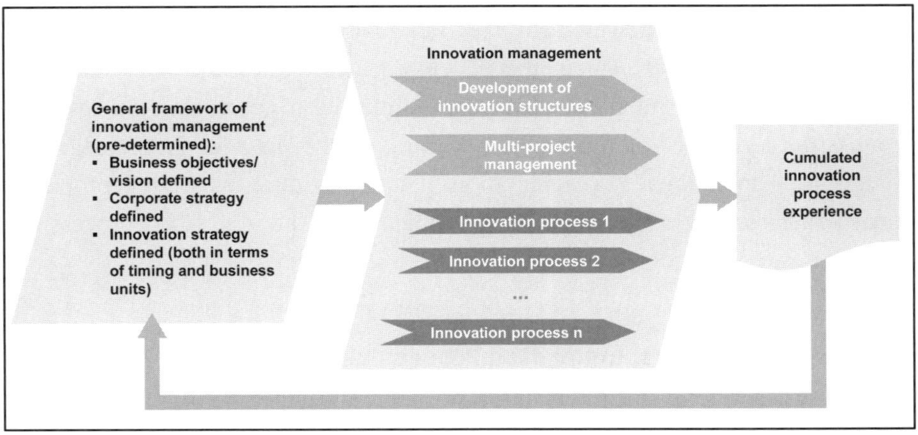

Figure 2: Innovation Management Framework

The surrounding conditions of innovation management are defined by company's strategy and its determinants, the companies' goal or vision. The innovation strategy derived from the enterprise's strategy determines the timing of innovation behavior, the specific fields of search (e.g., concentration on certain business segments or the exclusive search for cost-cutting innovations) as well as the priorities. The development of innovation structures includes the establishment and advancement of the innovation management concept within the operational und organizational structure. Thus, this function will be of low priority, once the balanced state is reached. It is much more important, however, for the establishment of the innovation management and for the accomplishment of structural changes. Multi-project management aims at the central coordination and control of the innovation project portfolio. In addition to resolving resource conflicts, multi-project management includes both the mentoring of members of the management of innovation projects and programs (e.g., training) and the provision of adequate instruments. Elements of innovation management which cannot directly be linked to a procedural understanding (such as an innovation culture of a company) are not shown in the framework.

The elements of innovation structure development and multi-project management could be led by a single operational unit such as the administrative departments of top management. Because of the high strategic importance of activities, corporate top management has to be involved not just in objectives and strategy definition, but also in the development of innovation structures and multi-project management. For individual innovation processes, top management is only involved as a promoter, backing up the plans of the managers in charge. The operational implementation of an individual innovation project is usually managed by a specialized project manager if the innovation process has reached a sufficient degree of formalization.

Modeling of Individual Innovation Processes

It was necessary to consider various requirements in the process of classification as well as the design of the major phases. Their specific implementation within the phase model is presented in Figure 3 in a comprehensive form.

Requirement	Consideration
Mapping feature	▪ Mapping of innovation processes at LSPs
Reduction feature	▪ Focusing on activities, allocation of actors, IT and management tools and on success factors
Pragmatic feature	▪ Purpose: analysis – No need for rollout – Preferably high level of detail – No need to map onto other models ▪ Focus on Fuzzy Front End – Separation of idea generation and selection from other activities – Precise partition within this major phase
Connectivity with innovation management system	▪ Integration into a processual framework of innovation management
Enabling learning with regard to processes	▪ Distinct activity IV.1: innovation process development
LSPs often being down-to-earth	▪ Composition of individual activities (esp. activities I.1, I.2, II.2 and IV)
Educational level of the employees often low	▪ Composition of individual activities (esp. activities I.1, I.2, II.1, II.2 and IV)
LSPs' organizations often decentral	▪ Distinct activity I.2: communicating ideas ▪ Distinct activity III.4: transfer to day-to-day business ▪ Separation of idea generation and selection from other activities ▪ Precise partition within this major phase ▪ Composition of individual activities (all activities in phases I and II)
High customer specificity of many LSPs	▪ Distinct activity III.4: transfer to day-to-day business ▪ Separation of idea generation and selection from other activities ▪ Composition of individual activities (all activities in phases I and II)
Service character of logistics	▪ Absence of activities aiming at producers of physical goods ▪ Composition of individual activities (esp. activities II.1 and II.3)

Figure 3: Consideration of Requirements for Classification of Phases

There are three somewhat classical phases in our model: "idea creation and selection," "innovation development" and "innovation realization." "Innovation process development" was added as a fourth major phase. This element aims at the institutionalization of the process improvement. It is important to reiterate that this includes learning concerning a single idealized innovation process. Learning in regard to the design of the whole innovation management system is illustrated in the framework in a cumulated way. This was deliberate; it is expected that changes concerning the innovation management as a whole will be so rare and of such top level responsibility that it would not make sense to establish a standard review process.

As a result of the consideration shown in Figure 3, there follows a sub-division of the four main phases of the idealized innovation process into twelve separate activities (Figure 4). The following sections present special results concerning every phase and activity in respect to the content, the allocation to central vs. decentral actors, to IT and management instruments and to success factors, as far as they are relevant. The expert panel found that LSPs consider the first, relatively creative part, of innovation processes as more challenging; the latter parts concerning standardized project management were of less concern. Hence, it makes more sense to concentrate on the earlier phases and activities.

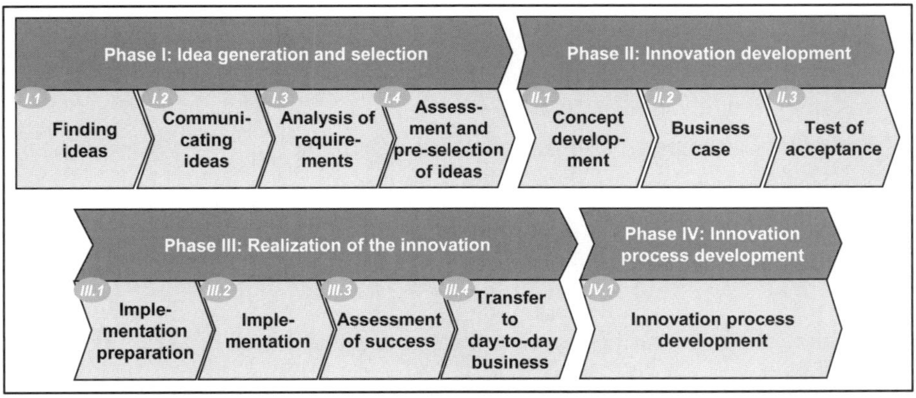

Figure 4: Idealized Model of an LSP-specific Innovation Process

Garcia and Calantone (2002, p. 112) state that "the innovation process is iterative in nature." Nevertheless it is practical to present models in a quasi-linear way. The user must simply be aware of potential multiple loops or even of chains of activities. In addition, innovation processes can be terminated without being explicitly mentioned in the model.

Concerning the identified success factors of innovation management at LSPs, the panel of experts generally accepted a number of success factors commonly identified in management and innovation research:

- Top management commitment
  (with respect to innovation strategy)
- Content-related focusing
  (with respect to innovation strategy)
- Pro-active search for ideas
  (with respect to innovation strategy)
- Standardization
  (with respect to innovation strategy)
- Central coordination unit
  (relating to innovation structure)
- Hierarchical control of innovation projects
  (relating to innovation structure)
- Interdisciplinary team structure
  (relating to innovation structure)
- Obligatory structuring of flow and decision-making
  (concerning the innovation process)
- Integrity of processes
  (concerning the innovation process)

In addition, there are further success factors, most of which relate to more than one activity, but not to the whole process:

- Stimulation of creativity, e.g., by using interdisciplinary working groups or by allowing a certain degree of freedom
  (primarily within the phase of idea generation and selection)
- Gathering of feedback from business partners
  (especially within the phase of idea generation and selection)
- Transparency of market and competition
  (especially concerning the major phases of idea generation and selection as well as innovation development)
- Utilization of key performance indicators, e.g., concerning customer satisfaction
  (especially for the individual activities of assessment of success and transfer to day-to-day business)
- Adequate HR policy in regard to recruiting and training
  (independent of phases or activity)

Phase I – Idea Generation and Selection

The aim of this phase is the creation and selection of innovation ideas. Emphasis should lie on the provision of methods and on a cultural environment allowing freedom for new approaches. Tight guidelines or standardized process flows are considered to be counterproductive.

Activity I.1: Finding Ideas

First of all, an idea is something new. Hence, the search for ideas is highly unpredictable and demands considerable creativity. Therefore, this activity can neither be structured nor divided into several steps, which eventually lead to the ultimate goal: a new idea. The reason for the search for new ideas can be either instrumental or functional. A new instrument is targeted to increase the efficiency of services by using new processes, tools and technologies. A new function includes a completely new service. Therefore, a new function aims to satisfy both present and future needs. Both approaches may be used for commercial exploitation, in the sense of innovation definition.

As of this writing, the search for ideas is mostly decentralized and upon the request of customers. This is because of the decentralized structure of service provision and of decision-making processes, combined with high customer specificity. In these activities, all decentralized employed members of an organization are potential process owners. The cultural barrier of the LSPs, generally being down-to-earth, could, however, result in a reduced willingness of individual employees to search for new ideas. Their relative lack of education could limit their capability to do so. It becomes clear that the relatively low innovation capability of LSPs, compared to other industries, is not by chance but inherent in their origins. Thus, it follows that the innovation management of LSPs should pursue an adequate staffing policy (the aforementioned comprehensive success factors) which should pay special attention to choice and qualification of staff (capability) as well as motivation and incentives (willingness).

Customers often propose new ideas, usually when they need a solution to a specific operational problem (generally a new instrument for a known function). Usually the customer does not come up with an immediate solution but gives the impetus for one. From that point, the client takes on the role of judge, deciding on the propositions and solutions made by the LSP. Therefore, it is advantageous to involve the client in the evaluation of ideas for potential solutions at the earliest possible stage.

A proactive attitude to finding new ideas is rarely found in this industry. Such an attitude, however, can be supported by appropriate tools and techniques. Potentially successful techniques include creativity techniques, the lead customer concept (von Hippel, 1986), prospection or early detection mechanisms, market and competition observation as well as the classical intra-corporate suggestion system. The motivation of employees and an adequate incentive system belong to the short-term and the selection and training of staff belong to the long-term success factors.

Activity I.2: Communicating Ideas

A distinct activity concerning the communication of ideas is rarely found in established models of innovation development. It was included here in order to encounter the innovation barriers which LSPs face and to constitute an active management of forwarding ideas. The most important of those barriers is the fact that organization of management and operations is often decentralized; the large physical distance of the units makes it difficult to exchange ideas. Therefore it is irrelevant whether the idea should be shared between two decentralized employees or between a decentralized employee and a central one (e.g., an innovation manager). The customer-related search for ideas (customer specificity as the second typical barrier) generally leads to the development of ideas which aim only at solving a specific local problem (of a decentralized unit). If communication of ideas is neglected, synergies are wasted. In this case the employees would reinvent the same thing, just in different places of the organization, and waste a large amount of resources. It becomes clear that the active promotion and control of ideas is essential. This necessity is considered through a separate activity in this model.

The actual embodiment of communicating ideas is influenced by two other limitations. Many LSPs, being down-to-earth, makes it unlikely that the search for information will follow a pull-principle. This could happen by distributing information according to the push-principle. To this end, an innovation newsletter could be used to promote new ideas. Employees should be sensitized to the importance of promotion of (push) and active search for (pull) ideas, which could be facilitated by the use of incentives. For example, naming individual employees in the innovation newsletter can have a large motivational effect without committing any financial resources. Therein it is important to consider the employees' relative lack of education. Accordingly, news ideas should be self-explanatory and communication should use basic and easily comprehensible terms. The communication should accommodate these requirements (e.g. by a central communication expert).

The use of an intra-net application could assist in the search for existing solutions (pull-principle). Such a knowledge management system was strongly promoted by the expert panel. The innovation newsletter could be used as an instrument allowing for push-communication. Alternatively, an official form, similar to those used for an intra-company suggestion system, could be used.

In regard to the success factors, both the ability dimension and the willingness dimension have to be considered. Concerning the ability dimension, it is important to inform all employees about the instruments existing within the company. Hence, high willingness fosters a pleasant, informal atmosphere for the exchange of ideas and is therefore a key to success. An informal setting is important because many people perceive too much formalization as unpleasant. Given that communicating ideas, in most cases, leads to more work for the employees, a relaxed environment might increase their willingness to do so. Thus, formalized exchange of opinions and experiences should be restricted to the experts.

Detached from individual innovation activities of LSPs, the creation of competitive, customer-specific products through the combination of modules and standardized components is a major success factor. The identification and selection of components capable of being used as a standard or at least being transferable to one specific case, is anything but trivial. It follows that the promotion of ideas must allow ready identification of the potential of components to be re-used in a different context.

Activity I.3: Analysis of Requirements

The analysis of requirements preliminarily to Activity I.4 (the assessment and pre-selection of ideas), ensures a well-founded basis for the ensuing steps. The completeness of documentation of requirements can be ensured by formalizing the flow and by putting strict guidelines on the output. It has to be ensured – as far as the market is not a quasi-anonymous one – that the customers' requirements are met. There is an increase of formalization within this phase, restricting the participants of an innovation process in certain ways.

Activities which control and monitor the analysis of requirements should be carried out centrally. Some of these activities are the development of output targets, the examination of the content of the documentation and its checking for completeness. The owner of the process should ideally be from the area of controlling or a manager from the innovation unit.

The actual analysis usually starts off with a brainstorming session about the possible requirements of the solution. This may happen in centralized and decentralized organizations. With rising customer specificity it becomes increasingly important to include the client in the process of defining the specifications. It is therefore important to differentiate "must-do" from "nice-to-have." This could be especially difficult, because the customers tend to assign every pleasant option to the "must-do" list, without being prepared to pay extra for it. Not every suggestion of the client should be considered, in order to avoid his pitfall. It is rather advisable to present a final documentation for the customer's examination.

When the analysis of requirements relies on a market or competitive analysis (for example, the development of non-specific solutions) it could be helpful to invite the participation of external consultants. Otherwise the analysis should be assigned to employees from the operational areas of the company. The centralization of these activities is generally of no advantage, unless the LSP has detailed knowledge about the customers' future business into which the solution is to be integrated. Given the decentralized structure and high customer specificity of most LSPs, this will hardly be the case.

The central finding concerning the tools and methods applied is the awareness that processes, review cycles, and formats of reports have to be formalized in order to ensure completeness of the analysis of requirements and to guarantee the quality of data. Correct interpretation of the customers' demands as well as industry and method-specific expertise of employees, determining the accuracy of the analysis, are the keys to success.

Activity I.4: Assessment and Pre-Selection of Ideas

At various phase transitions, management should define milestones with kill-or-go-decisions according to the stage-gate process model (Cooper and Kleinschmidt, 1993; Cooper et al., 2002a; Cooper et al., 2002b). The right time to take such a decision would be just before a substantial rise in the input of resources. Such a rise happens whenever the innovation development phase is entered. Therefore it seems reasonable to review existing ideas and to give formal approval for the development of the concept, where applicable.

A first screening of the ideas by internal experts can be conducted either in a centralized or decentralized manner. When a project is of high importance, the innovation proposition should be presented to the executive board (for internal innovation projects of LSPs) or to the customer, at the end of the phase. Concerning customer-specific solutions, the choice of alternatives is relatively simple: As the client is the one to pay for it, the preferred alternative of the customer will usually be followed down. The only criterion which the LSP must check is the transferability of the solution to different contexts and therefore customers. When there is a generic solution (with no customer specificity), several distinct alternatives should be evaluated and compared to each other.

The specific features of LSPs which are barriers to many activities influence assessment and pre-selection of ideas in such a way that the operational units of the company will have better (more up-to-date) information and data than the central units. This emphasizes the need for a formal kill-or-go-decision in advance of a substantial rise in resources. However, it does not influence the content of such a decision.

Cost-benefit analyses as well as scenario techniques are two of the classical tools and methods which best suit the assessment and pre-selection of ideas. The accuracy of the benefits evaluation of a solution for existing customers and the challenge to gain acceptance of all involved parties are the keys to success for this activity.

Phase II – Innovation Development

The following activities, assigned to the innovation development phase, are intended to refine the pre-selected ideas gathered in Phase I. These include the conceptual development and the refinement of ideas as well as rigid evaluation of feasibility and economical validity in order to ensure commercial success.

Therefore, process control is far more formalized and structured than in the first phase.

Activity II.1: Concept Development

The development of the concept is the first activity of innovation development. According to Bowers (1989), concept development of services can be compared to the construction of a prototype for physical goods: First of all, "policies, procedures and standards of performance" (Bowers, 1989, p. 19) including a structural description of the services' outline, should be developed. Analogous to this, individual specifications of individual dimensions (such as service, process, etc.) should be established. These steps are followed by the combination of all factors to an overall specification. Consequently, the refinement of content of the drafted and pre-checked ideas is the nucleus of concept development.

Concept development firstly consists of checking the set-up and the evaluation of resources needed for provision of the service. Generally, it is more applicable to perform these tasks decentralized. During the process of idea refinement, models consisting of individual solution modules may be constructed and tested in respect to performance and their chances of success. The main advantage of such testing is the feedback, allowing rapid modification of the concept. The resulting concept should be fixed at a certain time by a centrally communicated design freeze. This restricts the requests for modifications at a later point which usually results in additional costs. At this point, all involved parties must agree on the proposed specifications.

In the case of a customer-specific solution it is advisable that the actual customer act as a "final authority." This could happen by means of a test example of the model which is developed. It should precisely meet customer expectations, because discrepancies are hard to correct and could damage the relationship with the customer, whereas positive divergence is neither honored nor paid for.

The specificity of LSPs is of less concern for the development of concepts. Decentralization and customer specificity, however, result in the responsibilities being installed locally. Tests of services are generally more challenging than tests of physical goods. This is further emphasized by a low educational level.

IT models and simulations supporting the evaluation of solutions and allowing for identification of weaknesses are appropriate tools within concept development. Key success factors are the acceptance of the design freeze by all participants (including the customer) in addition to the correct degree of detail of the original idea. Industry and the employees' methodological know-how are also relevant.

Activity II.2: Business Case

A business case represents a formalized examination of the commercial potential of a developed model. The scope of this examination depends on the importance of the innovation project and varies from basic financial considerations to sophisticated business plans stating targets, strategy and so on. Central development efforts independent of any specific customer request should have business cases going beyond an individual customer's demand. Rather, they should incorporate a market study. Most decentralized project efforts are covered by a single large customer's high demand and therefore guarantee a sufficient level of profit. In this case the business case is reduced to a simple calculation of the tender.

Again, it is the business unit which has to use its advantage of up-to-date data. Thus, it should generate and compile the planning data. A central unit assumes the role of a supervisor, checking on the comprehensiveness and plausibility of the information. The process owners of a business case within a business unit would typically be the operational managers in charge of the implementation of the innovation. In a central project, these would be managers holding an innovation management function either in the area of controlling (management accounting) or in some administrative department.

Besides the LSP characteristics already discussed (decentralization and customer specificity) two other characteristics are relevant: educational level and LSPs being down-to-earth. Both are expressed by an aversion to formalized processes.

Hence, the guidelines of conducting a business case should not be too strict in order to avoid employees' resistance to the use of this tool.

Comprehensive market reports were described as among the least useful tools. Simpler tools like spreadsheets are used more intensively. Key to success within this phase is the realistic judgment of the situation. To ensure this, management accountants might act as critical counterparts of the managers conducting the assessment.

Activity II.3: Test of Acceptance

In advance of the realization of a project (accomplished in Phase 3), it is necessary to check on the acceptance of the affected party, usually the client. Insofar as there is only one client, requiring one specific solution, he has to be quite closely involved, especially because it is within his scope to make a kill-or-go-decision. The acceptance of the actual solution by the customer should ideally be pre-determined by the requirements analyzed earlier. If the solution development is not aiming at one specific customer, the concept should be tested with a customer in real life or involve customers in confidential discussions to pre-test the solution. This activity involves an iterative approach to discussions with the participants and of subsequent modifications of the concept. This loop should be run through until the customer satisfaction has reached an acceptable level.

Decentralization and customer specificity of the LSP again result in the usage of dispersed process owners. The agents could be people who are in contact with customers, employees within operations or the innovation project manager.

The only tool to be used for this activity is an adequate IT-testing environment. The key success factor is the embedding of all participants, in order to ensure that all arguments and positions are represented.

Phase III – Realization of the Innovation

Planning and implementation of the innovation as well as its transfer into operational reality (day-to-day business) is accomplished in Phase 3. This phase is structured like a project. It conforms to the general definition of a project, because of its time restriction, its non-recurring execution and its continuous evolution (Project Management Institute, 2004). Therefore, this phase should be controlled by standard processes of project management. Also, this underlines the opinions of the experts (who are very experienced in project management) on promoting the convergence of innovation and project management in the later phases of innovation development.

The process control follows well-defined steps and is therefore of a rather formal character.

Activity III.1: Implementation Preparation

The preparation of the implementation is the first task in realizing the concept. It includes the initiation of the implementation projects as well as the creation of an implementation road map. The involvement of stakeholders, especially clients, is critical for success.

Project initiation includes the preparation of the site of implementation (usually at the customer's site) and the training of users and support staff. A collective kick-off meeting of all participants is recommended. The implementation road map, which must subsequently be compiled, should contain the final definition of content and scope and standard project plans (such as a project structure plan, time and resource allocation plans and quality, staff and communication plans). The projected values within the project plan represent the basis for the assessment of success accomplished in Activity III.3. The individual design depends on the content and scope of the innovation to be created. All substantial activities have to be executed locally.

Planning can be accomplished by either by project management software or by standard forms. The key success factors identified are orientation along project management standards, staff management and training, and detailed preparation. In the case of general, non-specific solutions, it is important to have a well-functioning communication channel to the market.

Activity III.2: Implementation

Implementation activities involve the actual execution, the management of the implementation and project controlling. Execution of activities should ideally be anticipated, so that the implementation management and project controlling are engaged in coordination and not in planning.

During the implementation the targets that were set in the project management plan must constantly be revised. This is necessary in order to apply corrective measures in case of a deviation. Special attention has to be paid to all transition points of responsibility across company borders. This is the case when responsibility for a process is transferred from the LSP to the client and vice versa.

Again, standard tools of project management and reporting are amongst the instruments to be used for these tasks. Concerning the transition of responsibilities, pre-determined procedures should take place. Key to success at this stage is excellent project management.

Activity III.3: Assessment of Success

The final activity concerning the innovation as a novelty is an assessment of its success, including a formal acceptance of the innovation outcome. Upon completion of these activities, the innovation outcome has to be designed in accordance with the specifications. Other central factors in assessing the success could be customer surveys and analyses of customer satisfaction. Again, most steps would be conducted as decentralized efforts.

For products developed centrally, the customer should have access to formal process reviews in order to receive feedback in real time. When modifications are needed, they could be done right away. Help desk support and remote diagnostics tools could be used, in addition to operational and financial controlling instruments. Key success factors are rigorous preparation and the effectiveness of all review processes and tools.

Activity III.4: Transfer to Day-to-Day Business

The implementation project ends with the transfer into day-to-day business. Quite often, a (first) implementation without a preliminary assessment of ideas, development of concept, business case takes place. In this case a distinct transfer, prior to the day-to-day application, is obsolete. The transfer could rather be characterized as a rollout. As soon as a solution is implemented at one customer's it must be examined if it can be made available to other customers (and if so, which modifications would be necessary). Apart from centrally organized coordination, all activities will be accomplished peripherally.

A highly successful product (service) innovation can be marketed by a case study. Other ways for structuring customer feedback and improvements would be through reporting tools.

Prerequisites for success include the quality of modifications after an assessment and the quality of management after the implementation to reach the customer's acceptance. In the case of a transfer by rollout there are different success factors: Modularity and standardization of the innovation as well as intra-company transparency of accomplished innovations.

Phase IV – Innovation Process Development

The concept of organizational learning is now widely accepted in management science (e.g., Argyris and Schön, 1978; Easterby-Smith, 1997; Visser, 2007). Of particular importance is that the usefulness of post-project reviews has been demonstrated with regard to R&D projects (von Zedtwitz, 2002). Hence, the aim of the final phase is the evaluation and improvement of the innovation methodology and procedure during the innovation process. It has both formal aspects (e.g., process documentation and assessment of applicability as "Best Practice") and creative elements (e.g., creation of specific improvement recommendations).

The innovation development process is important and distinctive. Every participant ought to be given the opportunity to suggest ways to improve the process flow. This is of particular importance for LSPs who, due to their often decentralized organizations, are in a high danger of letting experience pass unnoticed. In addition, standardization of reviews helps to handle potential barrier effects from LSPs often being down-to-earth and having a low educational level. Naturally, passing on experience only helps if it is used an input for following processes. Hence, it is advisable to invite project managers to participate in the reviews of other projects (von Zedtwitz, 2002).

The expert panel recommended the following procedure. First, the innovation process should be described in writing. This activity has an internal purpose, i.e. as an inspiration for future innovations. The second purpose is external marketing. Additionally, it is advisable to invite distinct project participants to the debriefing and write a report about the contents of the meeting. The outcome (improvement proposition) can be used for future innovation development processes. At this point we want to repeat that the process of continuous learning is incorporated into the innovation management framework in a cumulative form.

We recommend undertaking post-project reviews at the very end of an innovation process. Only then is it possible to give the reviews the attention and time that they deserve. For processes terminated before a transfer to day-to-day business we suggest scheduling post-project reviews as soon as significant resources are invested and as soon as there is a division of labor. That should be the case every time Phase 2 (innovation development) had been entered.

Looking at the methods, there are two reasonable instruments in place: formal processes for gathering feedback and improvement proposals, and the standards to take them into account. Further, a comparative analysis of target and actual process values could prove effective.

Key success factors in Phase 4 are a pronounced feedback culture and sufficient allocation of time and resources.

6.4 Summary

Our research clearly shows that LSPs need to adopt industry-specific innovation management instruments and concepts. The frequently observable "down-to-earth character," combined with a relatively low level of education of their employees and their often decentralized organizational structure require the adoption of industry-specific innovation processes. Furthermore, the strong influence of individual customers and the intangible character of logistical services have to be considered.

Based on theoretical deliberations and action research with industry experts, an audit tool for innovation processes of LSPs was developed which takes into account the typical features of logistics services. Innovation managers at LSPs are offered an instrument which allows them to analyze bottlenecks within their companies and to suggest improvement measures.

We showed that the features of distinct activities as well as the tools and success factors vary throughout the innovation development. Figure 5 depicts an overall summary.

The allocation of the vast majority of tasks is handled decentralized. Some of the central activities are more supportive (e.g. the ones in the activities "finding ideas" and "communicating ideas"), whereas others are more supervisory (e.g. the ones in the activities "analysis of requirements" and "business case").

Additionally, it became obvious throughout the discussion of the distinct activities that later on, innovation process management can be handled with existing project management concepts and instruments. This is not the case for relatively early activities. As LSPs tend to handle project management very well, a managerial implication is that management effort focus on the early activities of innovation development.

Concerning future research, we suggest supplementing the actions research approach with case studies and extending the current process-oriented view (which we applied not only to the innovation processes but to all of LSP's innovation management) to a systems perspective.

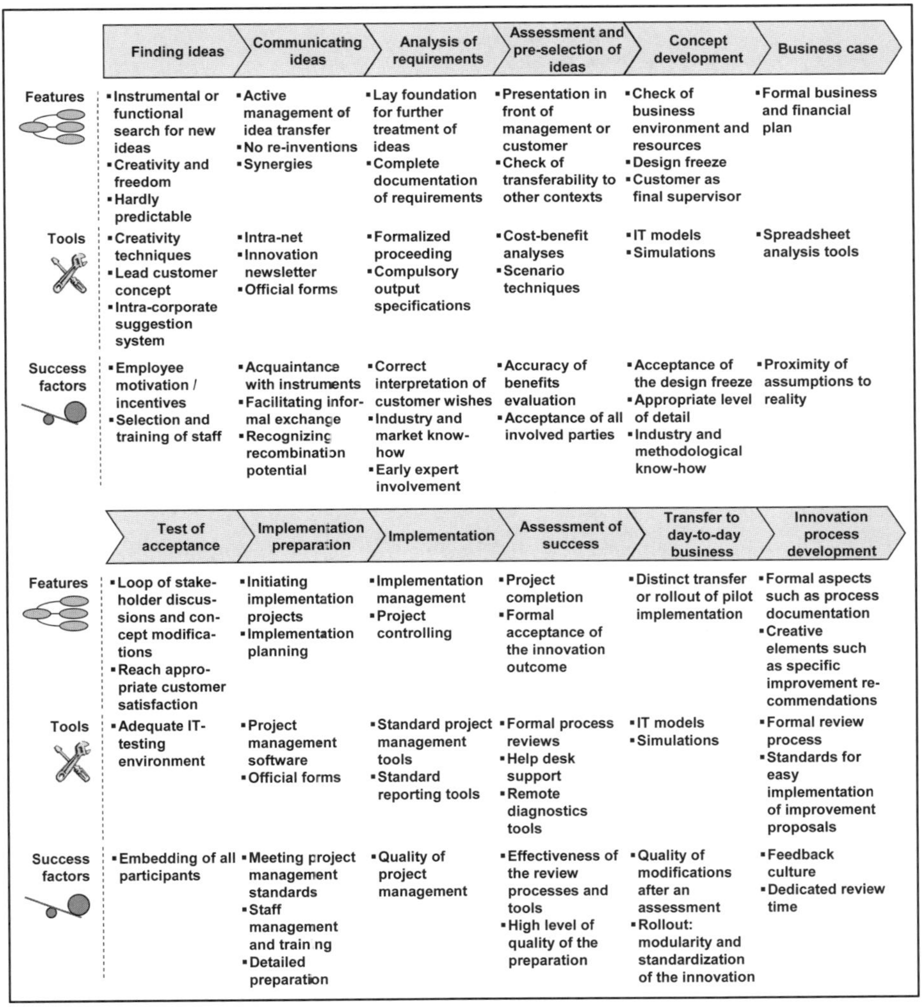

Figure 5: Audit Tool

References

Adams, R.; Bessant, J.; Phelps, R. (2006). Innovation Management Measurement: A Review, *International Journal of Management Reviews*, 8 (1), 21-47.

Argyris, C.; Schön, D.A. (1978). *Organizational Learning: A Theory of Action Perspective*, Reading, MA: Addison-Wesley.

Bichou, K.; Gray, R. (2005). A Logistics and Supply Chain Approach to Seaport Efficiency - An Inquiry Based on Action Research Methodology, in: H. Kotzab; S. Seuring; M. Müller; G. Reiner (eds.). *Research Methodologies in Supply Chain Management*, Heidelberg, 413-428.

Bowers, M. (1989). Developing New Services. Improving the Process Makes it Better, *Journal of Services Marketing*, 3 (1), 15-20.

Cooper, R.G.; Edgett, S.J.; Kleinschmidt, E.J. (2002a). Optimizing the Stage-Gate Process. What Best-Practice Companies Do (I), *Research Technology Management*, 45 (5), 21-23.

Cooper, R.G.; Edgett, S.J.; Kleinschmidt, E.J. (2002b). Optimizing the Stage-Gate Process. What Best-Practice Companies Do (II), *Research Technology Management*, 45 (6), 43-48.

Cooper, R.G.; Kleinschmidt, E.J. (1993). Stage Gate Systems for New Product Success, *Marketing Management*, 1 (4), 137-147.

Easterby-Smith, M. (1997). Disciplines of Organizational Learning: Contributions and Critiques, *Human Relations*, 50 (9), 1085-1113.

Garcia, R.; Calantone, R. (2002). A Critical Look at Technological Innovation Typology and Innovativeness Terminology: A Literature Review, *Journal of Product Innovation Management*, 19 (2), 110-132.

Koplin, J. (2005). Integrating Environmental and Social Standards into Supply Chain Management - An Action Research Project, in: H. Kotzab; S. Seuring; M. Müller; G. Reiner (eds.). *Research Methodologies in Supply Chain Management*, Heidelberg, 381-396.

Kühne, T. (2005). *What is a Model?, Dagstuhl Seminar 04101*, Dagstuhl. Available: http://drops.dagstuhl.de/portals/index.php?semnr=04101.

Middel, R.; Brennan, l.; Coghlan, D.; Coughlan, P. (2005). The Application of Action Learning and Action Research in Collaborative Improvement within the Extended Manufacturing Enterprise, in: H. Kotzab; S. Seuring; M. Müller; G. Reiner (eds.). *Research Methodologies in Supply Chain Management*, Heidelberg, 365-380.

Müller, M. (2005). Action Research in Supply Chain Management - An Introduction, In: H. Kotzab; S. Seuring; M. Müller; G. Reiner (eds.). Research Methodologies in Supply Chain Management, Heidelberg, 349-364.

Prockl, G. (2005). Supply-Chain Diagnostics to Confront Theory and Practice - Re-Questioning the Core of Supply Chain Management, In: H. Kotzab; S. Seuring; M. Müller; G. Reiner (eds.). *Research Methodologies in Supply Chain Management*, Heidelberg, 397-412.

Project Management Institute (2004). *A Guide to the Project Management Body of Knowledge (ANSI Standard 99-001-2004)*, Newton Square, PA: PMI Publishing Division.

van de Ven, A.H. (1986). Central Problems in the Management of Innovation, *Management Science*, 32 (5), 590-607.

Visser, M. (2007). Deutero-learning in Organizations: A Review and Reformulation, *Academy of Management Review*, 32 (2), 659-667.

von Hippel, E. (1986). Lead Users: A Source of Novel Product Concepts, *Management Science*, 32 (7), 791-805.

von Zedtwitz, M. (2002). Organizational Learning through Post-Project Reviews in R&D, *R&D Management*, 32 (3), 255-268.

Wagner, S.M. (2007). Logistics Industry: The Future Belongs to Innovative Logisticians, *International Transport Journal*, 69 (35/36), 14-15.

Westbrook, R. (1995). Action Research: A New Paradigm for Research in Production and Operations Management, *International Journal of Operations & Production Management*, 15 (12), 6-20.

Chapter 7

Logistics Innovation Processes: Potential Pitfalls and How to Prevent Them – An LSP Perspective

Britta Gammelgaard

Abstract

The increase in globalization emphasizes the importance of logistics solutions to connect ever more geographically dispersed production units with customers and end-users. Time and space have become essential parameters in global competition. New and innovative approaches to logistics have therefore become imperative in order to increase the speed of material flows. Yet the focus on logistics innovation has so far been limited. Firstly, the industry had to establish worldwide networks and alliances as the physical basis for unhindered material flows around the world. However, the time has come for logistics service providers (LSPs) to work with logistics innovation. But logistics innovation does not happen by itself. Management attention and organizational resources have to be dedicated to these essential processes. Even then, there is no guarantee of success. With Flint et al.'s (2005) Logistics Innovation Process Model as the point of departure, this chapter identifies potential pitfalls in logistics innovation processes and explains how such pitfalls may be avoided. A case of logistics innovation processes at Maersk Logistics outlines the most important pitfalls, and a theoretical analysis discusses the causes for their emergence, finishing with a recommendation on how they may be avoided. It is hoped that this will assist LSP's in preventing such pitfalls from inhibiting innovation among LSPs.

7 Logistics Innovation Processes: Potential Pitfalls and How to Prevent Them – An LSP Perspective

7.1 Introduction

This chapter describes potential pitfalls in logistics innovation processes. The first question that comes to mind is: Why logistics innovation – and why now? By answering this question we will see the importance of exposing and discussing the pitfalls of logistics innovation processes.

Both the logistics knowledge field and the logistics industry have been quite late in thinking about innovation, innovation processes and innovative solutions. Innovation has been reserved for physical products and product development. But innovation of logistics processes takes place all the time, even though quite anonymously as what it is: innovations. Yet, some of the most interesting organizational innovations in recent management history have something to do with logistics and supply chain management; the logistics principle of flow is essential in these innovations. For example, Toyota, Dell Computers and others have developed entirely new business models based on logistics principles (Franks, 2000; Hubbs, 2006) and their successes have been much talked about in the trade magazines and in the scholarly literature.

But the term "logistics" is also closely connected to the logistics and transportation industry. And even though not very visible in the general management literature and discussions, this industry has been very busy in the last few years taking care of the "physical internet" (The Economist, 2006). In other words, this industry has taken care of the materialization of visionary management ideas of globalization; not least with the help of McLean's standardized container usable for intermodal transportation in 1956 (Mayo and Nohria, 2005). In order to establish this "physical world wide web," the industry has had to create global networks in order to secure smooth and streamlined product and material flows around the world in the shortest possible time so that the end customer hardly realizes that his or her new purchase has been through a global manufacturing process. As a consequence, much effort in the industry has been dedicated to building strategic alliances and mergers. Now, the industry has consolidated itself sufficiently to start thinking of innovation. The constant development of new information technology and physical infrastructure in economically developed parts of the world also supports this trend. Last but not least, their customers are now ready for it as their supply chains have become more widespread and complex due to the intensive globalization processes and search for cost reductions.

All in all, with globalization as the driver of industry consolidation, information technology and infrastructure, the time has come for logistics innovation processes as one of the next important management issues. But as in all other innovation processes, there are potential pitfalls in logistics innovation processes. In order to highlight these pitfalls, I first present the point of departure: the Logistics Innovation Process Model (Flint et al., 2005). From there on, a case is presented from which central pitfalls are deduced. Each of these pitfalls is then analyzed and discussed from various theoretical perspectives. The aim of this discussion is to better understand the pitfalls, and thereby avoid them.

7.2 The Logistics Innovation Process Model

In the section above, Toyota's and Dell's business models were presented as examples of break-through logistics innovations. Following Rogers (1995, p. 11) in Flint et al. (2005), innovation is defined as "*... an idea, practice, or object that is perceived as new by an individual or other unit of adoption.*" The implication of this definition is, according again to Flint et al. (2005, p. 114), that logistics innovation is "*... any logistics related service from the basic to the complex that is seen as new and helpful to a particular focal audience.*" This means that in order to talk about a logistics innovation we do not have to talk about innovations new to the world such as, for example, the RFID tag. The customer's perception of newness is what counts. This definition therefore emphasizes the customer in the innovation process, but it also includes internal process innovations that result in new customer services that the customer *per se* has not been involved in creating. One such example is e-business firms that reduce delivery time "behind the lines," and thereby improve market attractiveness. Process innovations that have a purely internal focus are not included in this definition.

The discussion of logistics innovation processes and their pitfalls is based on the Logistics Innovation Process Model below. This model's starting point is a focal firm – for example the logistics service provider (LSP) – but the central point of the logistics innovation process is the interaction between the focal firm and a customer. The efforts of the processes are directed towards creating value for the customer first and the supply chain next.

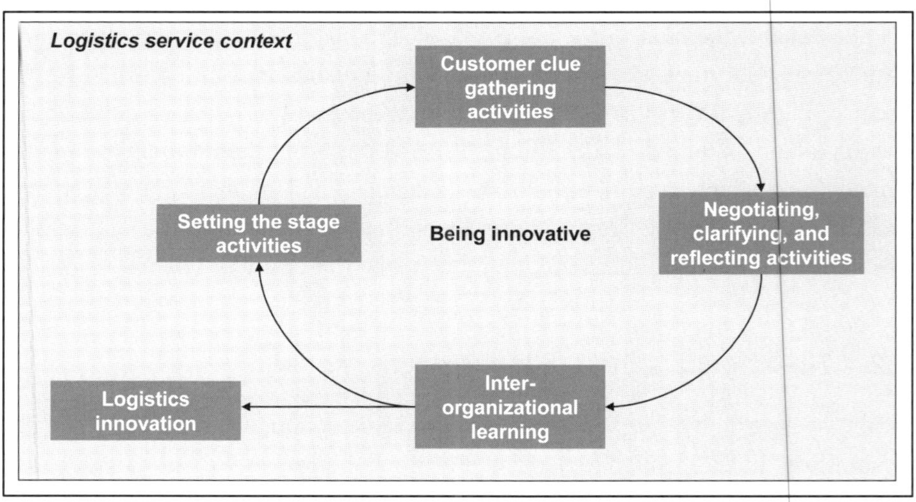

Figure 1: The Logistics Innovation Process Model (Source: Flint et al., 2005)

The Innovation Process Model takes its point of departure in "Setting the Stage Activities," which means that somebody in the focal organization is creating a framework in which the processes take place. In the Kraft case referred to in Kelly's "Ten Faces of Innovation" (2005) as "The Collaborator," this is equivalent to the setting up of a meeting between Kraft's supply chain people (supplier) and Safeway (customer), people who barely knew each other. Often logistics and supply chain employees do not actually meet with customer counterparts in person, but the Kraft supply chain manager Ron Volpe thought otherwise, and was willing to take the risk of letting the customer formulate his or her wishes with respect to the Kraft supply system. From Kraft's supply chain point of view, the risk is that so many and such overwhelming wishes are formulated that they would not be able to fulfill them, thereby leaving the customer unsatisfied. But Volpe was willing to go into a brainstorming session with Safeway, a session that opened up to all kinds of ideas. In the model, the brainstorming session was a "Customer Clue Gathering Activity." From there on the brainstorming ideas were clarified, reflected on and negotiated between the relationship partners, as illustrated in the next step, "Negotiating, Clarifying, and Reflecting Activities." In this process, the parties learned about each other's systems and needs ("Inter-Organizational Learning"), and thereby found new and innovative ways of organizing the supply chain. The result was a "Logistics Innovation" that created substantial economic benefits for the relationship. Among the results were an improved VMI (vendor managed inventory) system and a cross-docking arrangement, both minimizing the risk of out-of-stock situations and leading to shorter lead-time in general. Inter-organizational learning had taken place, learning that to some extent is relationship-specific, but most likely also contains generic parts usable in other relationships. Inter-organizational learning is the last process element before the process results in a logistics innovation or starts all over again. Successful or not, the process may be repeated as long as the partners like, or when one or both parties think there may be relevant issues for improvement, innovation and learning.

Besides making the logistics innovation process visible, this model provides an understanding of logistics innovation and the LSP's role in that process. This understanding lies in the activity of negotiation and reflection activities where the supplier and the customer meet. According to the process approach highlighted in the model, this is based on the idea of innovation as strategic reflexivity proposed by Sundbo and Fuglsang (2002, p.1). They argue that: "*Innovations have to be negotiated. Firms cannot merely adapt to markets as stable environments, they need to be more reflexive and strategic. By reflexive we mean that firms must take into consideration the changing manifestations of their actions on the market and society. By strategic we mean that firms must consciously negotiate their role in the division of labor and in relation to customers.*"

These ideas describe in a very precise way the experience of the firms in the case study. Mere adaptation to markets or customer requirements is not a feasible approach to logistics innovation. The LSPs themselves have to contribute knowledge and ideas to the process. Furthermore, they have – individually and with their customers – to consider the impact of their ideas and innovations on more than just the dyadic relationship. For transportation and logistics this process element is essential. Transportation and logistics create value by delivering the products and materials to the customer at the right place and time, in the right quantity and quality. Efficiency and effectiveness are central in logistics management. There is no need for prolonged reflection on the positive economic impact of logistics efficiency and effectiveness. But to an increasing degree, LSP's now have to include the impact of their services on the external environment, an aspect that has formerly not been central to supply chain thinking (New, 1997). More and more customers now want information on, for example, the amount of CO2 emissions as a consequence of transportation of their goods. This means that not only must LSPs reflect on their impact on the environment, they must also reduce it.

The last part of the citation above points to the negotiation of the role in the division of labor in relation to customers, and hence the supply chain. This element captures two of the most interesting logistics innovations in recent years, namely Vendor Managed Inventory (VMI) and Third Party Logistics (3PL). In VMI, the supplier manages the customer's inventory where it was formerly unthinkable to leave the responsibility of a vital business process to another firm. And in 3PL agreements, more kinds of logistics activities (e. g. final packaging) are transferred to logistics firms. These innovations mirror the developments in the global economy and management thinking: the attention to core competences. This is more than just a smart idea; for many firms, focusing is necessary in order to meet the demands of growing markets and market demands. By negotiation and changing the division of labor, customers are left room for improving their competitive edge. The downside of focusing on own core competencies is increased dependency on others. This dependency, however, can be a motivation for the LSPs, leaving them room for long-term contracts and stability, and hence productivity in their operations. But the handing over of vital operations – yet not core competencies – as logistics for many firms demands trust between the parties. This relational element plays an important part in the negotiation of logistics contracts and innovations.

The learning approach of this model is that of participation and interaction. Furthermore, the reflexive approach also indicates that double-loop learning may take place in successful negotiations. The participants question their own and their partners' mental models (i.e. anticipation of reality and logistics solutions). In short, the model suggests that all these process elements may lead to a logistics innovation.

7.3 Maersk Logistics: A Logistics Innovation Process Case

Maersk Logistics is a global LSP that offers customized, integrated supply chain solutions to customers around the world. Maersk Logistics has more than 7,000 employees in 200 offices in more than 90 countries. Maersk Logistics is a part of the A. P. Moller-Maersk group with Maersk Line, the largest container shipping firm in the world and one of the largest logistics firms in the world. Maersk Logistics also operates as consultants, and their solutions do not necessarily imply Maersk Line's or other sister companies' services. At the same time, the deep and detailed knowledge of global logistics networks and operations that the group possesses is a tremendous advantage to Maersk Logistics in designing new and applicable supply chain solutions.

Logistics Innovation at Maersk Logistics

In order to customize logistics solutions, close collaboration with customers in the innovation processes is essential. As the starting point, the customers are offered a "SupplyChain HealthCheck™," an analysis of the present state of their supply chain operations. This is done with an end-to-end IT-based supply chain simulation tool supported by a comprehensive consulting methodology. This tool was developed internally by the department *Supply Chain Development* and launched around 2000. Much to their surprise, they were first on the market with this comprehensive simulation tool, and they still lead a market where they compete with other LSPs and global consultancy houses. Their explanation for this is, again, their deep and practical knowledge of supply chain operations, not necessarily their IT expertise per se.

The *SupplyChain HealthCheck*™ approach consists of the following elements: 1) understanding the strategic priorities, 2) financial and operational benchmarking, 3) mapping current processes, 4) conducting gap analysis, 5) simulating future scenarios, and 6) planning solutions for implementation. Output from the analysis is AS-IS and TO-BE logistics costs, lead time, inventory levels, and efficiency in terms of capacity utilization through all steps from vendor/supplier to store/retail outlet. New solutions could involve transport mode optimization, new routes to market based on product characteristics, new order and warehousing procedures, and a revised distribution network. Part of the Maersk Logistics approach is tracking the value generated by the solutions after they have been in operation for 6-12 months.

Logistics Innovation Processes, a Customer-Oriented Approach

The *SupplyChain HealthCheck*™ is the point of departure for the logistics innovation processes that Maersk Logistics carries out with its strategic customers. The prerequisites for a good end result are a coherent project management approach and close interaction with all stakeholders within the customer organization. They will typically form a project organization consisting of *sponsors*, a *steering group* and a *project management team.* As these projects cut across boundaries, it is not unusual to see involvement of people from China, India, the Middle East, South East Asia, Europe and North America on a single project. Different habits, cultures, languages and time zones obviously complicate such projects, but they are a part of everyday life for a truly global company such as Maersk Logistics.

A new assignment will start with a request from the key client manager who is responsible for the customer. Then a workshop or kick-off meeting is held. The workshop may include some brainstorming but it is not a standard procedure per se. The role of sponsors at this time is to secure alignment of the project with the overall strategy of the customer. Furthermore, the involvement of the sponsors is necessary in order to anchor the project in the customer organization. Then the actual work starts with supply chain mapping, information gathering, benchmarking, simulation and analyses.

A project closes with a meeting with the sponsors, where new solutions are presented and discussed. Heavy emphasis is placed on the communication of the results of the analysis and the implementation plan to the customer organization. "Logistics as a concept is straightforward," says Global Head of Supply Chain Development at Maersk Logistics Erling J. Nielsen, "... ensuring effective implementation through tailored communication and stakeholder management is not." If the customer accepts the proposal, the implementation process starts with participation from both organizations.

A special department of the Maersk Logistics organization, *Supply Chain Development*, executes these analyses and projects. This department has grown from a few people to more than 70. The management premise is that of the *learning organization*. All projects must be documented and described as *case stories*, some of which are also used for marketing purposes. But for the organization itself, it is essential that all projects, procedures and methods are documented and shared via an internal *library*. All consultants must contribute to knowledge sharing and communication. This operational element is so important that it is a substantial part of the individual employee's KPI rating, which is based on the Balanced Scorecard. The internal learning process is accentuated by the "red teaming" process. This means that various parts of the department systematically review the projects, and ask provocative or critical questions in order to give feedback and enhance organizational learning.

7.4 Potential Pitfalls in the Logistics Innovation Processes

There are four pitfalls in particular in customer-oriented innovation processes at Maersk Logistics: lack of long-term dedication, lack of information sharing and mutual openness, lack of support from the customer's organization, and lack of openness towards the environment.

Lack of Long-Term Dedication

It is devastating for the innovation and supply chain process if the customer is not willing to make a long term commitment to the new solution and to the operators chosen for their integrated supply chain. Implementing new solutions takes time and is based on stability. If the customer is "shopping" and wants only the best prices in the short term, it may be done at the expense of larger savings that will materialize when taking a longer term view of the end-to-end supply chain. Maersk Logistics offers these innovative services mainly to its strategic customers who value long-term partnership.

Lack of Information Sharing and Mutual Openness

In order to come up with a reliable analysis of the situation "as is" and hence a benchmarking of the present solution, the LSP depends on the client's willingness to share information and openness to strategic priorities.

Lack of Support from the Customer's Organization

A project may apparently be well anchored in its own organization and yet still run into difficulties in the customer's organization. This may happen – again according to Maersk Logistics' experience – when the customer organization's performance measurement system is not aligned with logistics and supply chain priorities. One example of such misalignment is when a CEO is not interested in writing off obsolete stock on his balance sheet as it will impact his P&L negatively in the short-term. However, from an overall as well as from a supply chain perspective, this is the only right thing to do, since obsolete stock takes resources to manage and fills storage space which could be used for far better purposes. In a case when this situation occurred, it was not that the CEO did not understand the situation, but that he would be "punished" for not performing if he let the stock go due to this firm's performance management system. "And you cannot expect individuals to do something against their best personal interests" says Erling J. Nielsen, the Maersk Logistics general manager. "It is the task of the organization to secure a well-aligned KPI system that drives the right behavior."

Lack of Openness Towards the Environment

In order to stay attuned to the customers' present and future needs – whether they realize them or not – it is important to scan the environment for new trends, issues and thoughts.

7.5 Understanding the Pitfalls

Having identified the pitfalls in logistics innovation processes, avoiding them seems straightforward: be long-term oriented, share information, align your performance management system, and open yourself towards the environment. These recommendations do not look like "rocket science" and they should be easily implemented. But often they are not. In order to understand why this is so, it is necessary to look into the basic ideas or theories of logistics innovation. Only then can the reason behind the pitfalls be understood, and proper remedial action be taken; even better, the pitfalls can be avoided. The following section will place the pitfalls as experienced by Maersk Logistics in a theoretical perspective in order to understand the forces that are at play in logistics innovation processes.

The first three pitfalls are apparent in the supply chain literature. According to e.g. Mentzer et al. (2001), essential elements of Supply Chain Management are long-term relationships, and information sharing and cooperation, internally as well as externally. These propositions are consistent with the meta-theoretical foundation of logistics and supply chain management of system dynamics. Without mutual openness in the supply chain, supply chain management – the institutional framework of logistics – will not work, and neither will supply chain innovation processes. The fourth pitfall is openness to the external environment – for logistics innovation is easily recognizable from the open systems theory saying that the system needs input from the environments in order to survive and renew itself.

In short, a systems approach means that our mental model of, for example, a supply chain, consists of linked, interdependent entities. The interdependency means that there are stimuli and feedback processes between the entities in the system. When we – as logisticians – are trying to find and solve logistics problems, we look primarily at the links among the entities. In a physical supply chain, transportation and warehousing constitute such links. The Logistics Innovation Process Model is also built on systems thinking, although from a more recent systems approach, that of "second-order systems thinking" (Stacey, 2007). Inherent in this approach is the idea of systems, such as learning and innovation organizations, as "soft systems." This means that organizations function *as if* they were physical systems, but the system is a metaphor rather than a physical phenomenon. Even though this creates room for interpretations and subjectivity in the understanding of reality, it is still a systems theory with the same advantages and disadvantages (Stacey, 2007). Therefore in order to understand the pitfalls in logistics innovation processes, one must understand the pitfalls of logistics systems thinking.

Long-Term Relationships

Long-term relationships are a prerequisite for efficient logistics systems. Furthermore, long-term relationships promote willingness to share ideas, as both parties trust that new ideas will be treated in their mutual best interests; no new partners are jumping in and reaping the fruits of the efforts provided by the supply chain partners. Both parties also know that their efforts are not wasted by the partner ending the contract the next day. On the one hand, long-term relationships can secure stability and peace and concentration on work. On the other hand, these relationships slow innovation because both parties become too sure of each other, and eventually lose the incentive to improve and innovate. Furthermore, the partners may get to know each other too well. These phenomena are captured by the expression "the strength of weak ties" (Granovetter, 1983). "The strength of weak ties" means that it is through a peripheral network that one will gain inspiration and get acquainted with new people and new opportunities. More recently, this theory has been revitalized in the idea of diversity as a prerequisite for innovation. The diversity approach to innovation argues that if participants in innovation processes are becoming alike, the source of innovation vanishes because diversity generates new ideas (Johansson, 2004).

Information Sharing

Information sharing is essential in order to avoid the so-called bull-whip effect throughout the supply chain (Forrester, 1968). Furthermore, information-sharing is absolutely necessary if logistics flows are to be optimized across company borders, which is the aim of logistics and of supply chain management. No information sharing, no integration. And as logistics innovation processes should improve the logistics flows along the supply chain, information sharing is imperative. However, information sharing also may mean exposing oneself to power abuse by the next price negotiations. Furthermore, both parties need a set-up of business processes that handles such information sharing. This is time- and resource-consuming and may reduce the freedom of one or both parties to change partners.

Cooperation

Cooperation is the joint effort to integrate logistics flows internally and externally across organizational borders. A premise of this element of supply chain management is that of Supply Chain Orientation (SCO) (Mentzer et al., 2001), in which the supply chain parties must be willing to give up their own short term goals for the benefit of the supply chain as a whole. SCO is comparable to *holism* in systems theory. This means that if all parties work for the whole rather than themselves, it will create a synergistic effect where the efforts of the parties together exceed the sum of the efforts of the individual parties. The critical part of this proposition is the risk of the parties giving up their short term goals to the benefit of the supply chain as a whole (i.e. the supply chain partners). Are these partners willing to do the same after all, and how are the ultimate benefits shared? As for individual employees, rewards mean a lot to firms, and measuring and managing performance in a supply chain is far more complicated than internally in a firm with a hierarchical distribution of power. But even within a firm, the adaptation of a holistic approach means changes in the internal organization and hence in the distribution of power. This – in addition to basic lack of awareness of the driving forces – may very well prevent an optimal logistics process. This basic element in supply chain management thinking is perhaps the most critical, as it contradicts common sense and traditional economic theory to say that a party has to optimize its self interest.

Lack of Openness

Lack of openness to the environment may of course be caused by lack of resources for more than day-to-day operations. But another explanation may again be caused by more traditional systems thinking. If mental models of logistics systems and supply chain systems consist only of the partners in these systems, we "see" only these partners and nothing else. What is outside these systems appears irrelevant in reaching the goals set by the logistics and supply chain leadership. Furthermore, the system will have no influence on what is outside. This is a question of setting the boundaries of the system. As discussed before, this anticipation of the supply chain as a closed system may be dangerous to contemporary firms as customers are especially aware of the external ethics of global firms And even more importantly, lack of openness to the external environment precludes new ideas, new thinking and hence innovation. What the open systems approach does, however, is shed light on the system's relationship with its external environment. In fact, this approach suggests that the induction of "energy" from the outside is necessary to stay alive. Referring to the discussion on the strength of weak ties and diversity as a prerequisite for innovation, this may very well be the case in the longer run. With the intensifying global competition, it does not seem likely that any firm in the future will be able to survive in the longer run without continuously innovating its business.

7.6 Preventing the Pitfalls

It follows that the root of the potential pitfalls in logistics innovation lies in the way we are trained to think about logistics and supply chain management, namely that of efficiency and control facilitated by the Systemic Supply Chain approach. But when it comes to logistics innovation processes, other driving forces must come into play.

The *Logistics Innovation Process Model* is presented as a systems model, albeit one based on the idea of "soft systems." But the model may also be interpreted from the theoretical perspective of complexity thinking. In particular, the negotiating and reflection activities in the model indicate a step away from the systems approach without, however, being explicit about this. I will therefore elaborate on this alternative approach and outline some implications of this way of thinking for logistics innovation processes.

Complexity thinking consists of a large and growing array of approaches. The following will, however, take its point of departure from Stacey's theory of complex responsive processes (Stacey, 2007). Central to this theory is the manager's ability to control organizational processes and that of the existence of organizational boundaries within which managerial control take place. According to Stacey's line of thought, the ability to control is not a given, and hence there is no guarantee of predictable outcomes of management processes such as innovation. This contradicts main-stream organizational thinking of the firm's hierarchical structure as the safeguard of rational management processes and hence predictability (Johannessen, 2003). With the lack of importance of organizational structures, the boundaries of the organization, and hence also the supply chain, become irrelevant. Consequently, complexity thinkers would not apprehend a supply chain, but rather see complex, disorderly interaction processes between the entities of a supply chain. Nor will they see logistics innovations as a result of a controlled inter-organizational management process, but rather as something that may happen despite such processes. They will see logistics innovation processes in terms of cooperation and competition between the participating parties . This also means questioning the harmony of the systems approach; they rather see harmony and conflict as central driving forces of, for example, logistics innovation.

A "softer" version of complexity thinking will see logistics innovations as results of a deliberate decision of loosening control of organizational processes by balancing on "the edge of chaos" (McMillan, 2005). Either way, these ideas constitute a rejection of seeing the world as systems as we know them, open or closed. Rather, the world – and what we call supply chains – are perceived as consisting of actors that compete and collaborate at the same time, and where the outcome is uncontrollable and unpredictable. This means that lack of openness is not an issue in this line of thought; supply chains *are* open and constantly on the move as a result of the *negotiation, clarifying and reflecting activities* as depicted in the Logistics Innovation Process Model. These activities become the very center for understanding logistics innovation processes. Figure 2 depicts the differences between the two theoretical approaches.

	Approach	
Driving force	**Systemic supply chain**	**Complexity**
Control	▪ **Centralized control possible and necessary**	▪ **Centralized control not possible. Focus on control may damage subtle innovation processes**
Boundary setting	▪ **Essential for understanding and executing control**	▪ **Irrelevant as organizations are always in process**
Relationships	▪ **Harmony and cooperation for the benefit of the whole (holism)**	▪ **Harmony and conflict. Cooperation and competition for the benefit of strategic interest of entities**

Figure 2: Approaches to Understanding Logistics Processes

It therefore seems that the perception of a systemic supply chain is less suitable for logistics innovation processes. Does this mean that the ideas of systemic supply chain management should be abandoned? Hardly; this approach is far too successful to do so, but it was created to control something that has to be controlled: material flows. Rather, we should acknowledge its limitation when it comes to logistics innovation processes. When it comes to more subtle processes such as logistics innovation processes, we must first be aware that such processes may happen outside our system of control. Second, we must accept that something is and should be outside our control; and last, the organization must acknowledge these processes.

7.7 Conclusion

Logistics innovation is more relevant now that ever, not least because of the intensification of globalization. Although logisticians are accustomed to managing complex material flows around the world, managing innovation processes is perhaps not quite as familiar to them. Logistics innovation processes are delicate and hard to manage, because they involve more parties in the supply chain.

This chapter exposed the potential pitfalls in logistics innovation processes and suggested remedies. The case of Maersk Logistics revealed four potential pitfalls: lack of long-term relationships, information sharing, cooperation and openness. These pitfalls are also problems within systems theory so the discussion of how to prevent them took its point of departure in a discussion of the systemic supply chain approach. The conclusion was that while the systemic supply chain is well-suited for managing material flows, the emphasis on control, organizational boundaries and harmonic relationships is less well-suited for logistics innovation processes to emerge. And if this systemic supply chain approach is the only way to look at the organizational dynamics, innovation processes actually taking place in the organization may not even be perceived. So despite the indisputable success of the systemic supply chain approach, awareness of its limitations is recommended. The complexity approach is suggested as an alternative that emphasizes uncontrollable processes which, according to this theory, are the precondition for innovation. The recommendation of this chapter is therefore to change the mental model to create *awareness* of innovation processes as processes where cooperation and competition are constantly going on. In the Logistics Innovation Process Model, this would be constantly negotiating, clarifying and reflecting on the division of labor in the supply chain. This approach implies *acceptance* of the fact that not all processes shall or can be controlled by central management. Control is power, and overemphasis on power causes employees and business partners to submit to rules and systems rather than to contribute to new thinking. If participants act for their own benefit, if they compete and collaborate w at the same time conflict may erupt from time to time. Logistics innovation processes will only benefit from a more reflective attitude to control, for too much harmony and stability lead to stagnation of a business.

References

Flint, D.J.; Larsson, E.; Gammelgaard, B; Mentzer, J.T. (2005). Logistics Innovation: A Customer Value-Oriented Social Process, *Journal of Business Logistics*, 26 (2), 113-147.

Forrester, J.W. (1968). Industrial Dynamics – After the First Decade, *Management Science*, 14 (7), 398-415.

Franks, J. (2000). Supply Chain Innovation, *Work Study*, 49 (4), 152-155.

Granovetter, M. (1983). The Strength of Weak Ties: A Network Theory Revisited, *Sociological Theory*, 1 (2), 201-233.

Hubbs, J. (2006). Process Innovation. Toyota and Dell Profited from Speed. You can, too, *Manufacturing today*, September/October, 14-17.

Johansson, F. (2004). *The Medici Effect*. Cambridge, MA: Harvard Business School Press.

Johannessen, S. (2003). *An Explorative Study of Complexity Theory, Strategy and Change in Logistics Organizations*. Trondheim: PhD Thesis 91, NTNU.

Kelly, T. (2005). *The Ten Faces of Innovation*. New York: Currency Doubleday.

Mayo, A.J.; Nohria, N. (2005). The Truck Driver Who Reinvented Shipping. HBS Working Knowledge for Business Leaders, Oct. 3. Available: http://hbswk.hbs.edu/cgi-bin [Accessed 7 Aug. 2007].

McMillan, E. (2005). *Complexity, Organizations and Change*. London: Routledge.

Mentzer, J.T.; DeWitt, W.; Keebler, J.S.; Min, S.; Nix, N.W.; Smith, C.D.; Zacharia, Z.G. (2001). Defining Supply Chain Management, *Journal of Business Logistics*, 22 (2), 1-25.

New, S. (1997). The Scope of Supply Chain Management Research, *Supply Chain Management*, 2 (1), 15-22.

Rogers, E.M. (1995). *Diffusion of Innovation*, 4th Ed. New York: Free Press.

Stacey, R.D. (2007). *Strategic Management and Organisational Dynamics. The Challenge of Complexity*, Fifth Edition, Harlow: FT Prentice Hall.

Sundbo, J.; Fuglsang, L. (2002). Innovation as Strategic Reflexivity, in: Sundbo, J.; Fuglsang, L. (eds.), *Innovation as Strategic Reflexivity*, London: Routledge.

The Economist (2006). The physical internet – A survey of logistics. June 15th.

Chapter 8

Managing the Messy Process of Logistics Service Innovation

J. Rod Franklin

Abstract

The logistics service innovation process is a chaotic process. Standard models for the management of innovation activities in this type of an environment fail either because they are too rigid or because their ability to respond is restricted. In this paper the logistics service innovation process at logistics service providers (LSPs) is described and likened to "democratized innovation." Using the characteristics of this innovation type an approach to channeling innovation efforts based on management developed platforms and toolkits is proposed and an example is provided that demonstrates how this approach can be profitably implemented.

8 Managing the Messy Process of Logistics Service Innovation

8.1 Introduction

Logistics service innovation is a messy process. Unlike innovation in a product company or even in traditional service industries such as banking and financial services, innovation by logistics service providers (LSPs) occurs almost exclusively at the customer interface. LSPs, driven by the pace of customer demands, innovate on the fly to meet new or unique customer logistics requirements. This process results in the development of many new and innovative service offerings. These offerings, however, are developed in the high pressure environment of attempting to meet the immediate needs of a customer, not in the more controlled environment of a corporate development group.

Given the chaotic nature of the environment in which logistics service innovation occurs, it is not surprising that formal innovation processes are not widely employed by LSPs.[1] Formal phase gate reviews (Cooper, 1993) or technology funnel development activities (Wheelwright and Clark, 1992) applied to the multiple customer driven innovation efforts that occur simultaneously in a logistics operation slow down the delivery process. Because time is critical in customer-driven innovation, formal development processes are not employed. This does not mean that such processes are unsound or might not result in better logistics service innovations. What it implies is that such formal processes prove to be too time-consuming to be of practical use in assisting LSPs in managing their service innovation process.

If formal processes are of little use in the dynamic world of logistics service innovation, then how can LSPs ever hope to maximize their returns from field-based innovation? This question deserves far more detailed research so that LSPs can take action. In the absence of such research, however, this paper presents a practitioner's view on how one LSP's operations has approached this problem. The material presented is based on real life experience working for

LSPs attempting to leverage the many innovations that field personnel have generated for customers. The approach follows closely the concept of "democratized innovation" developed by Eric von Hippel (2005). However, unlike von Hippel, who assumes that democratized innovation employs users of a product or service as the primary innovators, the arguments there assume that innovations are co-produced by field-based LSP solutions teams and their customers, the end-users of the innovation.

8.2 Service Innovation in a Logistics Service Provider

Innovation within LSPs has traditionally been an ad hoc process. Because of the geographically distributed nature of logistics service provision, innovative service offerings tend to start out as unique service solutions developed for a single customer in a single region. These solutions can be classified as customer-driven innovations since they result directly from the perceived unique and immediate requirements of individual customers (von Hippel, 1988).

The solution development processes that the LSP's field-based personnel use to respond to these customer requests depend on the background of the response team, their familiarity with formal corporate guidelines and processes, and the nature of the customer request. However, all of these development processes are constrained by strict customer schedules. Time pressure forces the field team to make rapid decisions about the design that usually affect the generality of the innovation. Because of these design tradeoffs the service that is ultimately developed may only satisfy the customer for whom it was developed, not the broader customer market.

One "typical" logistics service innovation project at Kuehne + Nagel, a leading global LSP, demonstrates the many pressures, tradeoffs and resulting specialized solution that can develop from this process.[2] A telecommunications product provider was concerned about the sell through of their mobile telephones. Each month, product personnel responsible for the mobile telephone product lines received gross sales and turnover figures from their customers, the mobile telephone operating companies. Aggregate figures provided limited useful information because of their gross nature and timing. What the product managers needed were timely and focused geographic turnover numbers by mobile telephone type.

To obtain these more relevant figures the telecommunications product managers asked Kuehne + Nagel if it might be able, because of its current management of the company's distribution of mobile telephones to the mobile operating companies, to provide visibility to sell through data. Kuehne + Nagel's field-based solutions personnel sought an internal service system that might provide their customer with the information desired. One of the field engineers, upon hearing the requirements of the mobile telephone provider, concluded that Kuehne + Nagel's current Vendor Managed Inventory (VMI) software application, with slight modifications, would be appropriate in offering this service. The field operations team quickly modified the VMI application and implemented a customer acceptable solution based on this new application of the VMI tool set.

Once implemented, both the customer and Kuehne + Nagel's operations personnel noted that the VMI application provided capabilities beyond simply enhanced real time visibility. VMI services enabled by Kuehne + Nagel's VMI tools included the additional capability of performing collaborative planning and forecasting. The mobile telephone manufacturer's product personnel quickly grasped the potential implications of being able to better collaborate and plan with their customers. The product personnel leveraged these new capabilities to extend the original visibility service into a Collaborative Planning, Forecasting and Replenishment (CPFR) service offering.

The outcome of this interactive, distributed innovation process was a new service offering for the mobile telephone manufacturer that provided better sell through capabilities for their products and a potential new service offering for Kuehne + Nagel. However, because the new service was developed by Kuehne + Nagel's field team using existing tools and operating processes combined in an innovative way for a specific customer, the new service was not scalable or generic enough to be utilized for other customers. Significant re-engineering was required to take the service as developed for the original mobile telephone producer and generalize it into a CPFR offering for other customers of Kuehne + Nagel.

Several things can be discerned from this example of innovation within an LSP. First, logistics service innovation appears to arise not because of formal plans or processes, but because of pressure from customers to solve their perceived logistics problems. Second, time is of the essence in developing a response to the perceived problems. Third, innovations that arise from the customer's time-driven solution development process are usually composed of existing tools and processes combined in a novel manner. Because these innovations are co-produced with customers and LSP field solutions personnel working together, the combination of tools, processes and human factors may include external components from the customer or from external third parties. Chesbrough (2003) calls this type of innovation activity "open innovation" since solutions are developed from both internal and external inputs. Finally, these innovative services require significant rework and re-engineering for the LSP to capture their true value since their development does not occur with a general purpose or use in mind.

Although a single service innovation is described by the example, it has been my experience that almost all new service offerings in the LSP community arise in a similar manner. Logistics service innovations such as fourth-party logistics, lean logistics, order management services, merge in transit services, and many others all have arisen because of a demanding customer putting pressure on the LSP field support team. The solutions developed by the LSP field teams became the foundation of new service offerings from the LSP, but only after significant re-engineering was performed to generalize the focused innovation for sale to a broader range of customers. If this is the traditional way that logistics service innovation occurs within an LSP, then how can management within the LSP hope to control this messy process?

8.3 Democratized Innovation

LSPs generally operate highly distributed and autonomous networks for the delivery of logistics services. This organizational structure is the result of both the realities of modern transportation and distribution operations and the unique requirements of LSP customers. Freight movement requires that an LSP be able to manage a customer's goods from the source of production through the point of consumption. Given the increasingly international character of goods sourcing, successful logistics operators must have field operations in all parts of the world to ensure that they can support their customers. Management of the execution portion of logistics operations thus requires a highly distributed work force possessing local knowledge and empowered to make decisions in real time.

Traditional product focused innovation processes that assume a tightly controlled hierarchical business model fail to account for the highly distributed, dynamic innovation models found in LSPs. The formal innovation processes used by traditional product producing firms assume that innovation can be managed and controlled centrally. In the product producer model of innovation, central research and development groups attempt to identify customer needs and then develop solutions to these needs that can be leveraged by the firm to the best economic advantage. Innovations developed under this model are targeted at broad customer segments that have similar needs so that the largest potential market for the services can be served. Unique customer needs are, by definition, not the focus of these development efforts since a unique need does not define a sufficiently broad market to generate appropriate economic returns on the innovation investment (von Hippel and von Krogh, 2003).

Democratized innovation processes differ significantly from the traditional product producer innovation model. The democratized model of innovation assumes that innovation occurs in a highly distributed manner. End users of products or services are assumed to be actively involved in the innovation of the products or services that they use (von Hippel, 2005). This end user involvement in the innovation process arises because of numerous factors. The most important of the drivers of democratization are the increasing pressures of time, the greater availability of tools[3] that enable economic innovation at the point of use, and an increasing ability to access information necessary for innovation through modern communication channels such as the Internet.

The benefits of democratized innovation arise from the users' ability to develop exactly what they need instead of relying on someone else to develop something that might only imperfectly meet their needs. In addition, users may not have to rely on themselves to develop all that they need if they can leverage prior development work that is freely made available through other open innovation processes (Chesbrough, 2003).

The democratized model of innovation assumes that the manufacturer can still benefit from end user innovation efforts by acting as commercializer of the innovation. Since the end users are assumed to be interested only in solving their particular problem the manufacturer is free to take the unique solution developed by the end users, generalize it for a broader population and generate economic benefit from the commercialization of the solution. In effect the end users, by developing the innovation and piloting it within their own operations, short circuits the front end of the traditional innovation process and allows the manufacturer to go directly to commercialization without investing significant time in analysis and prototyping.

Applying the model of democratized innovation to LSPs requires modifying the "end user as innovator" concept. Innovation within an LSP's domain occurs, as discussed in the previous example, as an outcome of activities at the interface between the LSP's distributed field personnel and the customer/user. In the world of the LSP there is an ongoing dialogue at this interface that results in the co-production of an innovation by the user and the LSP field team. This cooperative innovation process operates like the independent end user in the standard democratized innovation model generating innovations that are focused on the particular needs of both the user and the local LSP team. These innovations, however, are not focused on the needs of the LSP or other customers. For the LSP to benefit from democratized innovation it must, like the manufacturer in a democratized innovation scenario, be able to commercialize the locally developed innovation.

8.4 Capitalizing on Democratization

The objective of any LSP is to create value and thus grow its business profitably. Innovations allow the LSP to differentiate its service offerings and generate greater profits. It is therefore extremely important for the LSP to capitalize on the innovations that its field personnel develop in response to customer needs. Unfortunately, given the democratized nature of LSP innovation activities and the lack of processes for controlling this distributed approach to innovation, capturing value from field innovation activities has not been easy for LSPs.

There are two ways for an LSP to capture value from its portfolio of distributed innovation activity. The first is the reactive approach. In this approach LSP field personnel innovate on behalf of a customer and the LSP, if it has the proper sensing tools, respond to the innovation by generalizing and commercializing it. The reactive approach to capitalizing on innovations is, in fact, the model generally followed by LSPs and reflects an opportunistic, but not strategic, response to customer requirements.

While the reactive model of commercialization predominates, it does not optimize the LSP's return on innovation activities. A reactive commercialization activity follows management's strategic intent only by happenstance. Reactive capitalization of innovations by LSP's is problematic since one can never be certain that the path that results from opportunistically commercializing a promising field developed innovation is the proper path for the LSP to follow. In addition, by simply reacting to innovations from the field, the LSP has no idea which innovation has the potential for greatest long term advantage. Selection, therefore, is at best opportunistic and, in certain instances, strategically and financially damaging.

Take as an example the decision by one LSP in the mid 1990s to commercialize an innovative order management solution that had been implemented for the delivery of training material for a franchise training company. The solution developed by this LSP's field team allowed the training company's franchise operators to view an online catalog of training material and company branded stationery products, place and track orders for these products online, and take delivery all in a closed loop process. The service took advantage of the then-emerging Internet as a mechanism for linking franchisees, the parent company and the warehouse where these products were stored in an integrated and real time manner.

Prior to implementing this customer-driven solution, the LSP had never participated in any online order management service. Its previous operations had, never offered any sort of high volume pick-and-pack operation. Ignoring this fact, and enamored by the potential profits that the business-to-consumer market promised, management of the LSP decided to commercialize the online order management service. Several years later, after spending several million dollars on re-engineering the field developed online order management system, the LSP abandoned its efforts. The LSP failed to attract any new customers for this service because of its lack of sales and organizational knowledge of the business-to-consumer market.

As this example indicates, simply reacting to an innovation opportunity developed through a democratized innovation process is not necessarily good. Management needs the ability to guide and channel the innovation activities in the field along pathways that are strategic for the company. Strategic intent, not opportunity and hope, is the key to maximizing an LSP's return on innovation.

The second approach to the capitalization of field innovations is the planned channeling of initiative. This approach requires the LSP to identify specific business platforms around which it wishes to build services and then develop toolkits that allow field personnel to rapidly assemble solutions on top of these platforms to the benefit of both customers and the LSP (von Hippel and Katz, 2002).[4] Since the toolkits provided by the LSP to its field personnel are not general purpose toolkits, but tools for building solutions on top of strategically developed platforms, they channel the innovation activities engaged in by the distributed field organization along pathways that the LSP can strategically exploit. However, since the toolkits allow the field teams and their customers developmental freedom within the platform framework, they facilitate innovation and value capture that are not present in the current ad hoc model.

The provision of toolkits to customers to create solutions for themselves based on a product manufacturer's platform is not a new practice (Ramirez, 1999). Software companies, for example, have provided toolkits (software development kits or SDKs) to developers for many years. In the physical product space, modular component designs have enabled the users of computers, music devices, and entertainment and gaming centers to customize their systems in highly innovative ways. In addition, these modular designs have enabled add-on equipment manufacturers to develop reliable and innovative extensions to these systems, thus enhancing the sell through of the original equipment manufacturer's "platform."

Unfortunately, within the logistics service market the strategic application of the platform and toolkit approach to channeling innovation has not been followed. Whether this fact is an outcome of customer demands, a lack of understanding, or a belief that such an approach cannot work in this market no strategic example of this approach has come to the attention of this author. However, a rudimentary form of this channeling activity has already been discussed here. In the mobile telephone manufacturer case the VMI application served as an unplanned platform and toolkit for the development of the CPFR system. Had management understood the issues of democratization of the innovation process and identified a base platform and toolkits for providing logistics services to the consumer electronics industry, a less costly and more widely deployable CPFR solution might have been developed. In addition, management could have assured itself that the distributed innovation efforts of its field teams were being focused down pathways defined by strategic intent rather than opportunity.

8.5 Piloting the Platform and Toolkit Model

Because LSPs have not employed strategically developed platforms and toolkits to channel field-based innovation activities one might wonder whether or not this approach would work in the LSP market. To test whether a true strategic channeling model might provide benefits management Kuehne + Nagel tested the model.

Kuehne + Nagel, like most LSPs, has grown from a regional to a global provider of services. Its field personnel maintain a significant degree of autonomy over their local activities and are often referred to within the organization as entrepreneurs. This reference clearly indicates management's perception, and expectation, that these individuals are to address customer issues in a pro-active manner without worrying too much about corporate approval. To ensure that control is maintained in the organization, a strong set of financial objectives and controls are used by the organization to focus the entrepreneurial actions of the field on opportunities that will return a positive monetary benefit to the organization. In addition, Kuehne + Nagel uses management policies and procedures to act as boundary controls for operations in the field. Field initiative to innovate within the constraints of the formal financial and budgetary system and management's policies is encouraged through Kuehne + Nagel's perpetuation of a corporate culture of entrepreneurship. The unique triangular control system employed by Kuehne + Nagel to control field activities is shown in Figure 1.

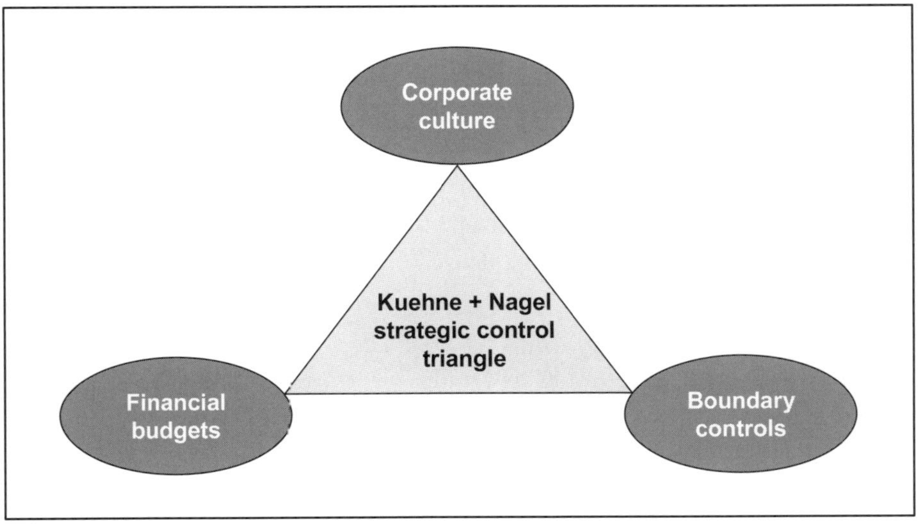

Figure 1: Kuehne + Nagel Innovation Management Control Triangle

Given the freedom that the management of Kuehne + Nagel has given its field organization to develop solutions for their customers, Kuehne + Nagel has developed a portfolio of services that is difficult to leverage to the benefit of the entire organization or its many customers. The concept of a platform and toolkit approach to channel innovation activities was presented as one way to gain broader benefits from local entrepreneurial activities to the management of Kuehne + Nagel. Although the concept was initially received by the management of Kuehne + Nagel with some skepticism, management decided to pilot in a real life situation. The decision to pilot the approach was made because of one senior executive's interest in the idea and the fact that it could be applied in a controlled manner in a new business that Kuehne + Nagel was about to enter.

This business field was the company's Lead Logistics Service (LLS). Because this was a new business field and required new systems and processes, management believed that employing a new business model for field activities could be implemented more successfully than in an existing business area where history and operating practices would interfere with any radically different approach to operations.

The LLS business, as defined by Kuehne + Nagel, concentrated on the management and not the execution of logistics delivery processes. This definition separated the LLS operation from the traditional business of Kuehne + Nagel in that it did not actually carry out any physical logistics functions; its role was the coordination and management of third parties (including other Kuehne + Nagel operating divisions) performing the physical activities. The primary functions of the LLS organization were:

- Network design, optimization, and setup
- Process development and planning
- Schedule development and management
- Contracting and contract management
- Financial auditing and reporting
- Continuous improvement
- General reporting and event management

Third-party service providers were to be managed by the LLS and perform work planned by the LLS team. The platform and toolkit approach to field innovation activities was believed by management of the LSP to be a good approach to facilitating a complex and highly variable new service offering.

After considerable analysis and discussion, management at Kuehne + Nagel decided on a software platform and service toolkit model (Figure 2).

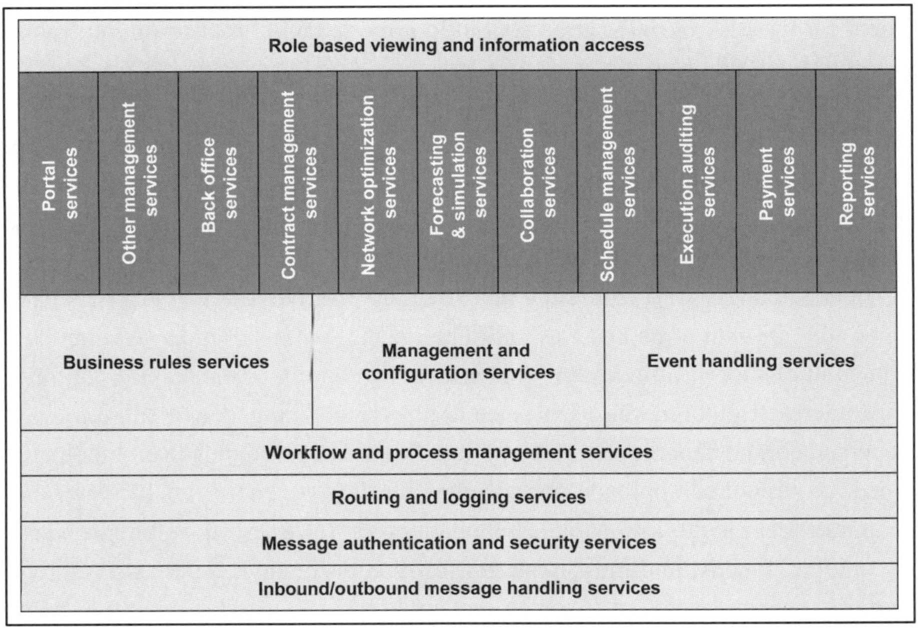

Figure 2: LSP Example of a Service Toolkit Model

The functional modules with vertical descriptions in the figure were to serve as the toolkits for field personnel. These toolkits could be used to construct customer specific services that would be "plugged" into the platform. Because the services would have been developed through the use of the toolkits they could be extended and leveraged for new customers or for expansion of existing customer services to new geographies.

The horizontal components (shown in Figure 2) were to provide common services for any custom solution developed through the use of the toolkits. Collectively, these services established a stable platform that provided an additional boundary for the development of new services. New services could be developed to the extent that the toolkits provided could be used for their development and to the extent that the common services enabled by the platform could support them. Through the use of the platform and toolkits Kuehne + Nagel's management hoped to channel field activities in a manner that met their strategic intent for the LLS business and that would provide higher returns on innovation than had been achieved in the past.

As with all new services, the development of the platform and toolkit concept was not without problems. Software and toolkit design issues delayed the implementation of the service offering by almost one year. However, once the software, toolkits and processes for implementing the service model had been debugged, the service was tested with extremely positive results. Field personnel were able to, within the design constraints of the toolkits, rapidly develop custom solutions for customers and implement them on top of the service platform. Customer satisfaction was high since the services being delivered were well documented and reliable. Service profitability was also higher than anticipated since the structured approach implemented through the use of the platform and toolkit concept led to lower implementation costs, fewer errors and higher quality outputs. Finally, and most important for this discussion, field-based innovation was now channeled in a manner that could be leveraged by Kuehne + Nagel for higher economic returns.

The positive results of using the platform and toolkit model at Kuehne + Nagel have led it to consider expanding this innovation model to additional business fields.

8.6 Summary

Innovation in any organization is difficult to control. Within the logistics services market, innovation is something that has traditionally "happened." Any thought of control or management has been secondary. This fact arises because of the manner in which innovation occurs in the highly distributed field and customer-focused business that is logistics service provision. Standard models for controlling innovation that use structured review and approval processes, although beneficial in other industries, have not proven to be of great use to LSPs. The inflexibility of these models and their need for central management does not fit with the democratized nature of LSP innovation activity. However, without some mechanism for channeling the innovative activities that do occur at the interface between LSP field teams and their customers, LSPs cannot hope to benefit to the fullest extent from the multiple innovations that their personnel develop.

This paper has described one approach, however, that appears to have merit and applicability to the prevailing model of innovation in an LSP. Innovation in an LSP occurs in a democratized fashion where the interaction between LSP field personnel and their customers is the locus of innovation activity. A democratized innovation environment can be strategically channeled by management through the use of a platform and toolkit innovation facilitation approach. This approach allows management to supply field personnel with toolkits that allow them, in conjunction with their customers, to rapidly build unique solutions that are framed by the capabilities of the toolkits and that fit on top of the strategic platforms of the LSP. An example of how this approach was implemented at one LSP indicates its potential for leveraging innovation and generating above average returns on the innovation efforts of field personnel.

The LSP industry and its challenges as a logistics intermediary between businesses have not received sufficient study. It is hoped that by describing how the platform and toolkit model has helped Kuehne + Nagel manage the innovation efforts in one of its divisions further research into the use of this model will be stimulated. Research into the democratized nature of innovation in logistics service innovation should lead to a better understanding of how to manage innovation in these companies. It is also hoped that further study of the LSP industry will generate new models of innovation and innovation management and thus advance this important field of research.

Notes

1 Logistics service innovation, as used in this paper, refers to the entire process of delivering a logistics service to a customer by an LSP. In this context it includes potentially new technologies (e.g., material handling technologies or information systems technologies), processes or human factor approaches. Logistics service innovation differs, therefore, from the much broader term of logistics innovation, which includes individual innovations in any discrete area of a logistics activity.

2 Examples discussed in this paper are disguised to protect the confidentiality of the customers that were involved in the activities.

3 The term "tools" is used in this paper to refer to mechanisms or software applications provided by the producing firm or third parties that allow the users of a product to customize it to fit their unique requirements. As such, software development kits, third-party add-in hardware for personal computers, third-party software that works on top of the MS Windows operating system, etc. are all tools that allow end users to customize relatively standard products for their unique requirements.

4 An example of a strategic platform is the personal computer. The personal computer as platform acts as a foundation for building custom solutions for users by allowing third party "plug and play" hardware and software to be integrated and combined in unique ways. The personal computer is a constraint to innovation by ensuring that all field-based customization must fit within the standards and capabilities defined by the personal computer design. By both constraining and facilitating innovation activities, the personal computer platform, along with the hardware and software tools developed to interoperate per the standards of the personal computer design, enables management to channel field initiatives along paths that have been defined a priori by management, while not requiring management to oversee every field-based innovation initiative that arises.

References

Chesbrough, H.W. (2003). *Open Innovation: The New Imperative for Crating and Profiting from Technology*, Cambridge, MA: Harvard Business School Press.

Cooper, R.G. (1993). *Winning at New Products: Accelerating the Process from Idea to Launch*, Reading, MA: Addison-Wesley.

Ramirez, R. (1999). Value Co-Production: Intellectual Origins and Implications for Practice and Research, *Strategic Management Journal*, 20 (1), 49-65.

von Hippel, E. (1988). *The Sources of Innovation*, New York: Oxford University Press.

von Hippel, E. (2005). *Democratizing Innovation*, Cambridge, MA: The MIT Press.

von Hippel, E.; Katz, R. (2002). Shifting Innovation to Users Via Toolkits, *Management Science*, 48 (7), 821-833.

von Hippel, E.; von Krogh, G. (2003). Open Source Software and the 'Private-Collective' Innovation Model: Issues for Organization Science, *Organization Science*, 14 (2), 209-223.

Wheelwright, SC.; Clark, K.B. (1992). *Revolutionizing Product Development: Quantum Leaps in Speed, Efficiency, and Quality*, New York: The Free Press.

Chapter 9

Is There a Role for Logistics Service Providers in New Product Development?

Zach G. Zacharia

Abstract

Researchers and practitioners have noted that logistics service providers (LSPs) have become more important in the economy as more companies focus on their core competencies and outsource everything else. However, a core activity in most firms is new product development and very few firms consider the benefits of incorporating LSPs in their new product development process. Through the use of several examples this chapter explores some ways in which LSPs can bring value to their client's new product development process.

9 Is There a Role for Logistics Service Providers in New Product Development?

9.1 Introduction

Logistics service providers (LSPs) have become more important as the business environment has changed with rise in global integration, shorter product life cycles, the push to reduce costs and the increased reliance on outsourcing functions and activities that are not part of the core competency of the firm. The LSP industry has been growing as companies demand more advanced logistics services (Hertz and Alfredsson, 2003).

There has been a trend to outsource new product development (NPD) as part of a process called "rapid distributed innovation" (Chesbrough, 2003). Outsourcing NPD can reduce development costs, shorten time to market, improve flexibility and gain access to specialized resources of external suppliers (Carson, 2007). In this chapter, NPD is a generic term that refers to new services and service innovations.

Despite the importance of LSPs, particularly in terms of the potential impact on effectiveness and performance in supply chains, few studies have analyzed the relationship between LSPs and new product development. LSPs have traditionally focused on providing services that are relatively easy to outsource such as warehousing, and transportation. Can LSPs be of value in strategic activities for the customer such as NPD? This chapter explores the effect of LSP involvement in NPD of their client firms.

9.2 The Importance of Innovation and Some Basic Typologies

Innovation is critical to the success of many firms, including providers of logistics services (Flint et al., 2005). In the broadest terms, innovation can be described as a subjective novelty which is open to a conscious management process and which aims at economic exploitation (chapter 1). A common way to distinguish between types of innovation is to characterize them as administrative or technological (Damanpour, 1991). Technological innovation pertains to products, services, and production process technology; it is related to basic activities and can concern either product or process (Damanpour and Evan, 1984). Administrative innovation involves organizational structure and administrative processes; it is indirectly related to the basic work activities of an organization and more directly related to its management (Damanpour and Evan, 1984). Even though an LSP can assist its client firms in both administrative and technological innovation, the LSP is likely to find greater familiarity and success associated with technological innovation.

LSP can focus on the process to introduce a new or improved product, process, or service into the marketplace (Betz, 1997). Lin (2006) defines a "new" product or service in that its cost is lower, its attributes are better, it now has attributes it never had before, or it had never existed in that market before. LSP can assist their client firms in two ways, by direct involvement in NPD process and/or any of the chain of activities that the firm must perform in order to offer the new product/service.

Another way to classify innovations is to consider them as either radical or incremental. Radical innovations are completely new to the marketplace in terms of product class or technology. All other innovations can be classified as incremental innovations such as new technology, new markets and product modification. Considering that the vast majority of successful innovations are incremental, this chapter will concentrate on incremental innovations.

A lot of research has focused on developing innovative products or services for the client, but is it possible to help the client in developing an innovation? The majority of the companies involved in outsourcing to LSPs want to reduce costs; however a few companies who outsource logistics activities want to improve the service to their customers. LSPs should target those companies to develop new products and new services.

9.3 How LSPs Create Value for their Customers

LSPs typically focus on ensuring a timely response to requests, on-time service delivery, accurate information storage and delivery, ability to solve problems, fulfillment of promises and assisting the clients in accomplishing their own objectives (Panayides, 2007). These goals identify the need for a close working relationship between LSPs and their clients to facilitate collaboration and more efficient exchange of information to be responsive to changing customer needs. LSPs would prefer to form client relationships for long term customers in order to reduce transaction costs as well as increase speed and flexibility in their business transactions. This is supported by the fact that LSPs are an integral part of their client's supply chain and are fundamental in serving the interests of the final consumer who could well be the customer of their clients (Panayides, 2007). This supports the concept of helping one's client meet the need of its customer. Figure 1 incorporates the view that LSP can become involved in the NPD product team and provide their input based on their understanding of both suppliers and customers to the client firm.

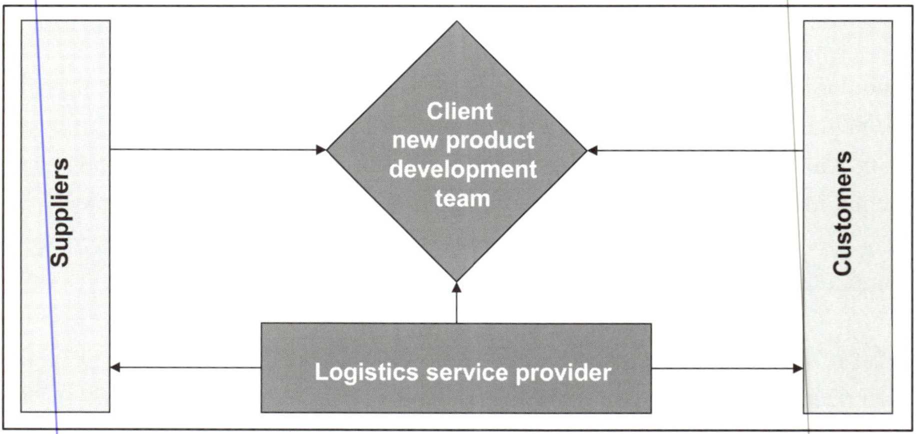

Figure 1: Logistics Service Provider NPD Model

Logistics innovation can arise from managers who seek to better understand unmet or poorly served customer needs and then develop new services that better meet those needs (Flint et al., 2005). Our research on LSP innovations led us to interview several firms in search of examples where LSP influenced the innovation process for a client firm. We asked LSPs to provide examples where they were directly involved in NPD projects that targeted their client's customer. It is also important to realize that many LSP actually have direct contact with a client customer purely through the handling of their logistics functions. Here is an example that illustrates the role of an LSP in developing a new innovative service to meet the needs of the LSP's client's customer.

For the sake of confidentiality the names have been disguised. A 3PL company called Alpha that specialized in the handling of high tech, highly sensitive shipments met with Beta, a client that delivered a broad range of goods to retail stores throughout the US. Beta a $50,000 a year customer of Alpha was satisfied with Alpha's ability to meet their requirements, but Alpha was convinced this business could be increased. Alpha sent a team to meet with Beta to discuss their customers and new opportunities for innovation. Beta invited Alpha to meet with their client Gamma to help in the delivery of a standard software program that was distributed to retail stores throughout the US. This software program is updated annually and there was strong evidence that new customers purchase these programs on how early the product is available in the store. Therefore the delivery of the product was very time-sensitive and the verification that the product was available in the sales floor was very important to Gamma. There had been instances in the past where the product was delivered on time according to Beta but not available in the sales floor.

Gamma hired temporary staff to visit retail stores to identify when the product appeared on the sales floor and this information was faxed back to Gamma headquarters. Alpha, after meeting with both Beta and Gamma, developed an online tool that showed transit information along with an interactive capability for field reps to access the system and report if the product was available on the sales floor. Gamma was highly satisfied with this new tool. Beta was satisfied in finding an innovative solution for their customer which led them to expand their account with Alpha from $50,000 to $2 million in just three years. This example illustrates the valuable role that LSPs can play in innovation and NPD, especially when developing a product for their client's customer.

A very important aspect of LSP innovation is to understand their customers and what those customers value. Customers select products, services, and providers of those products and services based in large part on how well they create value for them (Flint et al., 2005). Customers are concerned with the trade-off between functional, service, and relationship benefits, and monetary and non-monetary sacrifices as they relate to specific goals (Ulaga, 2003). LSPs create significant value because they have the potential to impact both the benefit and sacrifice aspects of customers' value perceptions (Flint et al., 2005). LSPs can understand what customers value indirectly through analyzing data such as obtained from sales reports. However, the more effective method of understating what a customer values is through direct interaction with the customer such as through face-to-face meetings (Flint et al., 2005). It is also important to note that customers value changes over time; this is a dynamic and iterative process that requires a close and strong relationship with the customer. For this process to be effective it is better to consider client meetings as an opportunity to determine if there are unmet needs the client *has with their customers*. Active searching for new opportunities with existing clients is a critical part of LSP innovation.

LSPs have also been an important component of business strategy and can provide a basis for a competitive profit edge (Olavarrieta and Ellinger, 1997; Richardson, 2005). LSPs can provide critical information on decisions about transportation modes, the most efficient transportation choice within the mode, the impact of packaging, and delivering the product in the desired condition. LSPs are an indispensable part of the economy, especially after transportation deregulation and improvements in information technology that have enabled companies to compete on the basis of competence in delivery speed, reliability, responsiveness, and low cost distribution.

LSPs also play a critical role in the launch of new products. For example, Scholastic Inc. (publisher of the Harry Potter books in the U.S.) collaborates with its LSPs, in the management of distribution operations to ensure the Harry Potter books go on sale at midnight throughout the U.S. on the same day. A key part of the success of this launch is to deliver Harry Potter books in a two-week time frame with no security breach. Several innovative services needed to be provided by the LSP which required active and direct involvement with the client and their customers in order for the product launch to be successful.

LSP can also assist NPD by understanding the perspective of the suppliers that provide products to the client. The common theme is the need to be willing to solve problems and develop innovative products that meet customer needs. To understand these needs, LSP should be willing to meet the client's suppliers. For example, an LSP was having problems with delivery performance for a client that needed finished vehicles delivered to individual dealers. The client was in a port city that received vehicles that arrived weekly on ships but arrival dates and product mix could vary. The suppliers to the client provided accessories that matched the number and type of vehicles arriving, two days prior to the vessel arrival date. Not having the right accessories resulted in vehicles not being assigned to the correct dealer, delaying shipments, adding significant costs and additional pressure on the LSP, leading to poor delivery performance. In this example the LSP met with the client's suppliers and discovered many of the problems were caused by the poor visibility of upcoming orders from the client's antiquated purchase order system. The LSP helped develop a new purchase order system using a web-based interface that provided the suppliers with automatic alerts for vessel arrival dates and the correct mix of accessories needed, as well as an expedited approval and payment process that improved supplier compliance. The client has a clear, real-time snapshot of their automotive accessory supply chain, suppliers have better compliance, and the LSP has better delivery performance. The LSP, by meeting with the suppliers, and by using their experience with multiple clients played a critical role in solving the purchase order problems with an innovative solution.

LSPs have been shown to play an important role in customer service (Sharma et al., 1995; Stank and Daugherty, 1998), product availability, time advantages, low cost distribution (Vega, 2004), and global transportation (Ritter et al., 2007). LSPs also interface with the client firm's manufacturing, marketing, and R&D departments to deal with procurement, inventory, warehousing, and distribution (Thuermer, 2007). By developing collaborative relationships with strategic client firms LSPs can also play a valuable role in a client's NPD process.

9.4 The Specific Benefits of LSP Involvement in NPD

The NPD process can also benefit from information provided by both suppliers and customers, which can be facilitated by an LSP. Supply chain management in the context of NPD can incorporate suppliers' and customers' ideas – again facilitated by LSPs. LSPs are more salient in firms which belong to industries that emphasize information technology and time-based competition (Zacharia and Mentzer, 2004). Firms where an LSP provides a competitive advantage will likely benefit from LSP involvement in NPD.

LSP involvement in NPD is defined as the act of integrating LSP representatives in the NPD decision making processes. Zacharia and Mentzer (2007) have noted the benefits of logistics involvement in NPD in terms of improved NPD project performance. Just as early supplier involvement in NPD (Clark, 1989; McGinnis and Vallopra, 1999; Mikkola and Skjoett-Larsen, 2003; Wagner and Hoegl, 2006) has an impact on NPD project performance (McDermott and Handfield, 2000; Song et al., 1997), LSP involvement should benefit NPD.

LSP involvement in NPD can be of benefit in many ways. When product life cycle times are short, as in fashion goods or electronic products like cell phones, logistics processes can shorten the time it takes to bring a new product to market (LaLonde and Powers 1993). Companies like Dell, Intel, and Sony – that use a modular product design – or automotive companies like Toyota, Ford, and General Motors – that use just-in-time manufacturing – benefit from early involvement of LSP in NPD. LSP feedback influences new product design and parts configuration, based upon reliability, serviceability, shipping, storage, and installation requirements (Huang and Mak, 2000; Meyers and Tucker, 1989).

LSP innovations do not have to be radical but could also be incremental. One of the LSP managers we interviewed was asked to help speed up their cross docking process. A simple solution was the addition of a colored piece of paper to the pallet as it is shrink-wrapped. The different colors represent the different distribution centers; this tells the exact destination of the pallet. This simple solution has accelerated cross docking and ensures that the pallets are going to the correct distribution centers.

The timing of LSPs' involvement likely affects the type of involvement in NPD. Costs incurred during the design stages of the NPD process may be no more than 8% of the total product development cost, but the decisions made in this stage determine as much as 60 - 80% of total NPD costs (Raia, 1993). Figure 2 depicts the many stages of a typical NPD project. It is important to note the earlier LSPs were involved, the easier it would be to make cost trade-off decisions among different designs.

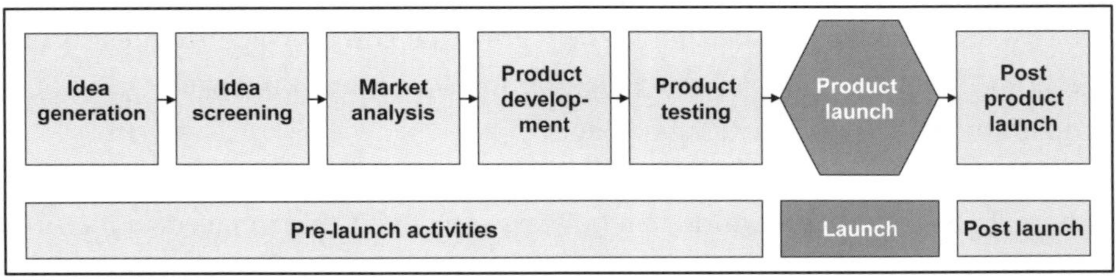

Figure 2: Typical NPD Stages

In the case of a popular automobile, the LSP noted that if the design team were able to reduce the length by two inches it would allow for 18 vehicles per railcar instead of 15 vehicles as in the proposed design. The design team developed a new length that met the reduction in size. That change dramatically reduced the costs of transportation on the order of several million dollars over the life of the vehicle. Reducing the length would have been impossible once the vehicle reached production, even if the LSP had been able to point out the cost savings. This suggests that the earlier LSPs become involved in the product development process the easier it is to make production cost versus logistics benefit tradeoff decisions.

A review of the NPD literature also identified many similarities among LSP involvement, logistics involvement and supplier involvement in the context of a NPD team. Birou and Fawcett (1994), researching supplier involvement in NPD, noted that supplier involvement is characterized by the timing and quality of involvement in the new product team. Zacharia and Mentzer (2007) also found support for the concept of logistics involvement in NPD. In the same manner, LSP involvement in NPD should lead to new services and new product designs that mitigate logistics costs and thereby improve NPD project performance and NPD logistics performance. Zacharia and Mentzer (2004) identified the increasing importance of logistics within the firm, the concept of logistics functional salience. Logistics involvement in NPD has led to improvements in product performance and logistics performance (Zacharia and Mentzer, 2007) further emphasizing the valuable role of LSP in NPD.

By being involved in the NPD process an LSP could make suggestions to reduce the cost and damage associated with delivering the new product. Another example from the interviews illustrates the role logistics played in ensuring the new products were delivered without damage.

Automobiles that are transported on rail cars used to be held in place with chains that were attached to the frame and ratcheted into place. Engineering redesigned the automobiles from having heavy solid frames to a monocoque construction (practically eliminating the frame) to reduce weight and save on fuel costs. Unfortunately, the LSP was not aware of this change, and by the time the automobiles reached the final destination, their frames were bent.

LSP's can also provide feedback on shipping conditions, such as exposure to water, air, dust, vibration, and temperature, which can affect the damage-free delivery and transport of the new product. Having an LSP involved in the NPD process should improve NPD logistics performance, especially during product delivery and transport. Another aspect of NPD logistics performance is to consider it as a component of NPD project performance. Improving NPD logistics performance in areas such as order fill rate or damage-free delivery should translate into more satisfied customers, which should directly improve the bottom line.

9.5 Managerial Implications

The primary managerial implication is that NPD project performance improves with LSP involvement in the NPD process. Companies frequently search for ways to improve NPD performance, and NPD is essential for the long-term survival for many firms. Much of the NPD literature suggests that companies still do a poor job of NPD and this chapter should be of great value for companies that do not currently involve LSPs in NPD. This chapter did not consider the type of new product or service; however, it is possible to speculate on the role of LSP involvement in NPD. If the shipping conditions, such as exposure to water, air, dust, vibration, and temperature, can affect the final product, there is value in having an LSP involved in NPD. When the physical characteristics of the product prevent the efficient utilization of cubic space, there is value in getting an LSP involved in the NPD process. If the new product requires special transportation, it is important to get an LSP involved in the NPD process. When the product launch becomes critical or there is a need to distribute products to a large number of retail locations in a very short time, LSP involvement in NPD is valuable. If the new product requires changes to the existing method of distribution, LSPs needs to be involved in the NPD process. If the cost of distributing the product is a significant component of the final cost of the product, then it is critical to have an LSP involved in the NPD process.

The second managerial implication is that LSP involvement in NPD improves NPD logistics performance. This suggests that companies concerned with the efficient distribution of new products, damage-free delivery, or those competing on the basis of logistics capabilities, benefit from LSP involvement in NPD. Given that a large proportion of NPD projects fail, any effort by the firm to improve performance is warranted. Since involvement of an LSP in NPD project teams improves NPD project performance and subsequent performance of logistics, firms should consider the strategic advantage this double performance enhancement brings to them.

The third managerial implication is that the timing of LSP involvement in the NPD process is important. Interview data supports the idea that LSP involvement in the NPD team led to designs that better incorporated logistics trade offs. This provides support for the notion that involving an LSP after the new product is completely developed is not as valuable as involving an LSP early in the NPD process. Companies developing new products where logistics issues are important clearly benefit from early LSP involvement in NPD.

The fourth managerial implication is that the type of LSP involvement in the NPD project impacts NPD project performance and NPD logistics performance. The quality of LSP involvement in terms of creativity, contribution, ideas generated, and ideas implemented, does matter. Therefore, having logistics personnel simply attend a few meetings of the NPD team might not directly translate into improvements in NPD project performance and NPD logistics performance. LSP involvement in NPD should focus on developing new creative ideas that can be implemented. The relationship of logistics with other members of the NPD team also affects the outcome of the project. The commitment and cooperation by the LSP to the NPD team are therefore important. Again, if the LSP is unable to significantly commit itself to and cooperate with the NPD team, there is little benefit in involving it in NPD. Both the quality and the type of relationship of the LSP involvement in NPD affect the outcome.

This chapter fills a gap in the literature with respect to relating LSP involvement in NPD and firm performance, and is one of the few to examine the relationship between LSP involvement in NPD, NPD project performance and NPD logistics performance. This research suggests there is strategic value in considering the role of LSP in NPD. As logistics functions are outsourced and as logistics capabilities play a role in the competitive nature of the firm, there is a clear benefit in involving LSP in NPD.

References

Afuah, A. (1999). Strategies to Turn Adversity into Profits, *MIT Sloan Management Review*, 40 (2), 99-109.

Betz, F. (1997). *Managing Technological Innovation: Competitive Advantage from Change*. New York: John Wiley & Sons, Inc.

Bienstock, C.C.; Mentzer, J.T.; Bird, M.M. (1997). Measuring Physical Distribution Service Quality, *Journal of the Academy of Marketing Science*, 25 (2), 35-44.

Birou, L.M.; Fawcett, S.E. (1994). Supplier Involvement in Integrated Product Development: A Comparison of US and Europe Practices, *International Journal of Physical Distribution and Logistics Management*, 24 (5), 4-14.

Carson, S.J. (2007). When to Give Up Control of Outsourced New Product Development, *Journal of Marketing*, 71 (1), 49-66.

Chesbrough, H.W. (2003). *Open Innovation: The New Imperative for Creating and Profiting from Technology*, Cambridge, MA: Harvard Business School Press.

Clark, K.B. (1989). Project Scope and Project Performance: The Effect of Parts Strategy and Supplier Involvement on Product Development, *Management Science*, 35 (10), 1247-1263.

Damanpour, F.; Evan, W.M. (1984). Organizational Innovation and Performance: The Problem of Organizational Lag, *Administrative Science Quarterly*, 29 (3), 392-409.

Damanpour, F. (1991). Organizational Innovation: A Meta-Analysis of Effects of Determinants and Moderators, *Academy of Management Journal*, 34 (3), 555-590.

Flint, D.J.; Larsson, E.; Gammelgaard, B.; Mentzer, J.T. (2005). Logistics Innovation: A Customer Value-Oriented Social Process, *Journal of Business Logistics*, 26 (1), 113-147.

Hertz, S.; Alfredsson, M. (2003). Strategic Development of Third Party Logistics Providers, *Industrial Marketing Management*, 32 (2), 139-149.

Huang, G.Q.; Mak, K.L. (2000). Modelling the Customer-Supplier Interface Over the World-Wide Web to Facilitate Early Supplier Involvement in New Product Development, *Proceedings of the Institution of Mechanical Engineers - Part B - Engineering Manufacture*, 214 (9), 759-769.

LaLonde, B.J.; Powers, R.F. (1993). Disintegration and ReIntegration: Logistics of the 21st Century, *International Journal of Logistics Management*, 4 (2), 1-12.

Lin, C.-Y. (2006) Influencing Factors on the Innovation in Logistics Technologies for Logistics Service Providers in Taiwan, *Journal of American Academy of Business*, 9 (2), 257-263.

McDermott, C.; Handfield, R. (2000). Concurrent Development and Strategic Outsourcing: Do the Rules Change in Breakthrough Innovation?, *Journal of High Technology Management Research*, 11 (1), 35-58.

McGinnis, M.A.; Vallopra, R.M. (1999). Purchasing and Supplier Involvement: Issues and Insights Regarding New Product Success, *Journal of Supply Chain Management*, 35 (4), 4-15.

Meyers, P.W.; Tucker, F.G. (1989). Defining Roles for Logistics During Routine and Radical Technological Innovation, *Journal of the Academy of Marketing Science*, 17 (1), 73-82.

Mikkola, J.H.; Skjoett-Larsen, T. (2003). Early Supplier Involvement: Implications for New Product Development Outsourcing and Supplier-Buyer Interdependence, *Global Journal of Flexible Systems Management*, 4 (4), 31-41.

Olavarrieta, S.; Ellinger, A.E. (1997). Resource-based Theory and Strategic Logistics Research, *International Journal of Physical Distribution & Logistics Management*, 27 (9/10), 559-587.

Panayides, P.M. (2007). The Impact of Organizational Learning on Relationship Orientation, Logistics Service Effectiveness and Performance, *Industrial Marketing Management*, 36 (1), 68-80.

Raia, E. (1993). The Design Team: Suppliers Move into the Line of Fire, *Purchasing*, 115 (8), 9-13.

Richardson, H.L. (2005). What are you Willing to Give Up?, *Logistics Today*, 46 (3), 27-29.

Ritter, L.; Barrett, J.M.; Wilson, R. (2007). Security First (Book Review), *Supply Chain Management Review*, 11 (1), 66.

Sharma, A.; Grewal D.; Levy, M. (1995). The Customer Satisfaction/Logistics Interface, *Journal of Business Logistics*, 16 (2), 1-21.

Stank, T.P.; Daugherty, P.J. (1998). Pulling Customers Closer Through Logistics Service, *Business Horizons*, 41 (5), 74-80.

Song, X.M.; Montoya-Weiss, M.M.; Schmidt, J. (1997). Antecedents and Consequences of Cross-Functional Cooperation: A Comparison of R&D, Manufacturing and Marketing Perspectives, *Journal of Product Innovation Management*, 14 (1), 35-47.

Thuermer, K.E. (2007). Where, Why and How Many?, *Logistics Management*, 46 (3), 47-49.

Ulaga, W. (2003). Capturing Value Creation in Business Relationships: A Customer Perspective, *Industrial Marketing Management*, 32 (8), 677-693.

Vega, R. (2004). Warehousing: Time to Think Outside the Box, *Logistics & Transport Focus*, 6 (2), 16-21.

Wagner, S.M.; Hoegl, M. (2006). Involving Suppliers in Product Development: Insights from R&D Directors and Project Managers, *Industrial Marketing Management*, 35(8), 936-943.

Zacharia, Z.G.; Mentzer, J.T. (2004). Logistics Salience in a Changing Environment, *Journal of Business Logistics*, 25 (1), 187-210.

Zacharia, Z.G.; Mentzer, J.T. (2007). The Role of Logistics Involvement in New Product Development, *Journal of Business Logistics*, 27 (1), 83-110.

Chapter 10

Implementing an Innovation Management System at a Logistics Service Provider – A Conceptual Guideline

Christian Busse and Stephan M. Wagner

Abstract

Implementing an innovation management system at a logistics service provider (LSP) is a challenging project. To support LSPs in this task we develop and present an implementation guideline in this chapter. Nine individual tasks within the implementation process are derived from a general innovation management framework. These tasks are assigned to the key actors, top and middle management. Presented in logical order, they can be clustered into three phases. With an additional planning and conviction phase at the beginning, there are four steps in total.

10 Implementing an Innovation Management System at a Logistics Service Provider – A Conceptual Guideline

10.1 Introduction

The best system design is useless if its implementation fails. That applies to innovation management systems as it applies to any other system, and it applies to logistics service providers (LSPs) as well as to other companies. We define implementation as the realization of solutions which exist in conceptual form and have to be transferred into concrete corporate action. Accordingly, in this chapter we will no longer consider the actual design of system elements to be implemented, but take them as a given instead.

We assume that it is not possible to give recommendations about *optimal* implementations due to the existence of contingency factors (e.g., Burns and Stalker, 1961). Since the specific requirements of LSPs and even their reasons for wishing to implement an innovation management system might vary, we focus on the matter at hand; we describe the tasks which have to be undertaken, the actors who might do so, and we provide an implementation roadmap. The value of our contribution lies in the novelty, as so far, there is no guideline for the implementation of an innovation management system at an LSP.

In this chapter, we deduce the main tasks in the implementation process and present them in chronological order. Those tasks will then be assigned to the key actors of the implementation process. The chapter concludes by deriving a four-phase implementation plan including support activities.

In the introductory chapter, innovation management was defined as follows: *the conscious configuration of the innovation system of a company or of a business unit.* This was mostly done to emphasize the fact that to the editors' understanding there is more to innovation management than merely the management of individual innovation processes. But what does an innovation management system look like?

10.2 Main Tasks of the Implementation Process

A number of components of innovation management systems have been listed throughout the book. Innovation management entails the definition of an innovation strategy and its alignment with corporate strategy, the design of organizational structures to foster innovation and to facilitate the management of individual innovation processes, the creation of a suitable innovation culture, the development of tools to support innovation processes and the management of individual innovation processes and of their interdependencies (Figure 1).

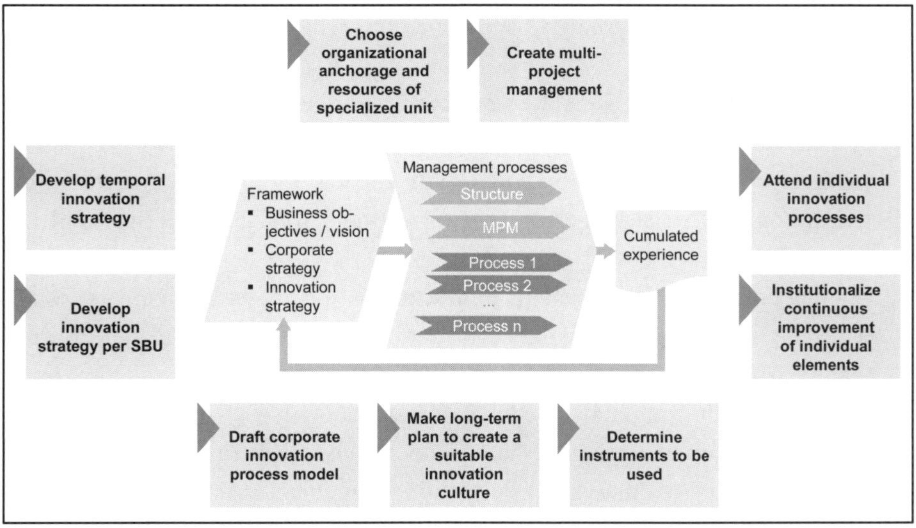

Figure 1: Implementation Task Determination through Innovation Management Framework

A company that has not implemented innovation management must complete nine major tasks before the stage of full implementation can be reached. (We assume that business objectives are defined and that a corporate strategy exists.)

1. In order to concretize the company's strategy, an innovation strategy has to be developed (Anthony et al., 2006). It has to interlock and support all other strategies without any interference. The temporal dimension of the innovation strategy includes the decision about innovation leadership, follower or deliberate imitation.
2. The concrete specification of the content of an innovation strategy should be organized by business units, either by aiming at revenue growth through product innovation within a very "young" unit (according to the product life cycle) or by aiming at revenue growth through process innovation within an "older" unit.
3. The decision of an LSP to implement an innovation management system automatically results in overall responsibility. This definitely includes the coordination, if not the planning of certain parts of the system. Therefore it has to be defined who should be assigned this role and which resources shall be reserved for it.
4. An innovation culture should be established. Such a culture is understood to provide an environment which has awareness of and fosters innovations that are independent of a specific agent, only ensured by the structure of the organization. Senior management plays a critical role in this (Zairi and Al-Mashari, 2005).
5. A procedural model to manage future innovation projects should be established. It should recognize the organizational structures and responsibilities that exist in reality. Differences among business models such as shipping or contract logistics might also be taken into account.
6. A multi-project management has to be created. This includes a strategic component such as innovation portfolio management and an operative component such as the management of resource conflicts between projects.
7. Furthermore, certain instruments have to be developed to support the pursuit of innovations. These could be management and/or information technology tools (e.g., knowledge management tools).
8. Furthermore, in today's dynamic environments, rules should be established to enable future modifications of tools and processes (e.g., Hyland et al., 2003; Soosay and Hyland, 2005; von Zedtwitz, 2002).

9. Finally, there should be rules about attending individual innovation processes. Roles for top and middle management might range from sponsorship to supervision to the actual project management. Plans should be made on how to support specific kinds of projects, as well.

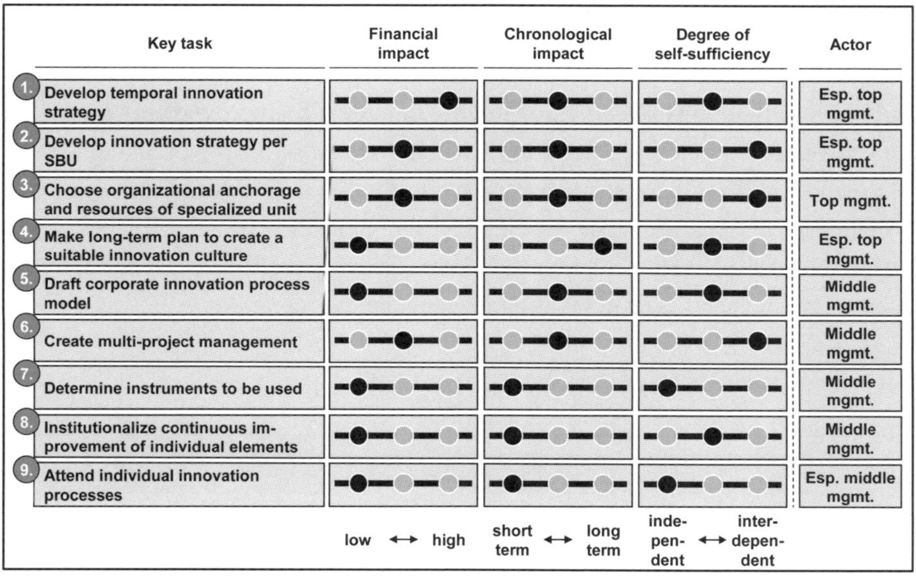

Figure 2: Implementation Task Allocation and Order

We distinguish two hierarchical levels of actors who can fulfill implementation tasks: top and middle management. The task allocation between these is based on three criteria: financial impact, chronological impact and degree of self-sufficiency. The greater the financial impact, the longer-term the chronological impact; the more interdependent with other tasks a given task is, the more likely it is to be executed by top rather than middle management. Both the evaluation of the individual criteria and the aggregation to a decision about an actor occurred, to the authors' best knowledge. The ordering of tasks is organized top-down, from top to middle management tasks, and from strategy to structure to tools to processes to continuous improvement and operative doing. (The resulting task allocation and order is shown in Figure 2.) It is worth noting that most of the individual assignments are somewhat interdependent, so that some backwards loops might be necessary in the implementation process.

10.3 Implementation Plan

Three distinct clusters can be formed within the chain of tasks. The first are the assignments which have to be performed by top executives, the second are initial implementation tasks, and the third are continuous (improvement) tasks. The latter two clusters are expected to fall within the responsibility of a single process owner from middle management. Due to the limited power of that actor and a certain level of resistance within the LSP's organization it may be necessary to supplement core implementation tasks with support work to persuade the employees of the benefits of innovation management.

The overall proceeding is depicted in Figure 3.

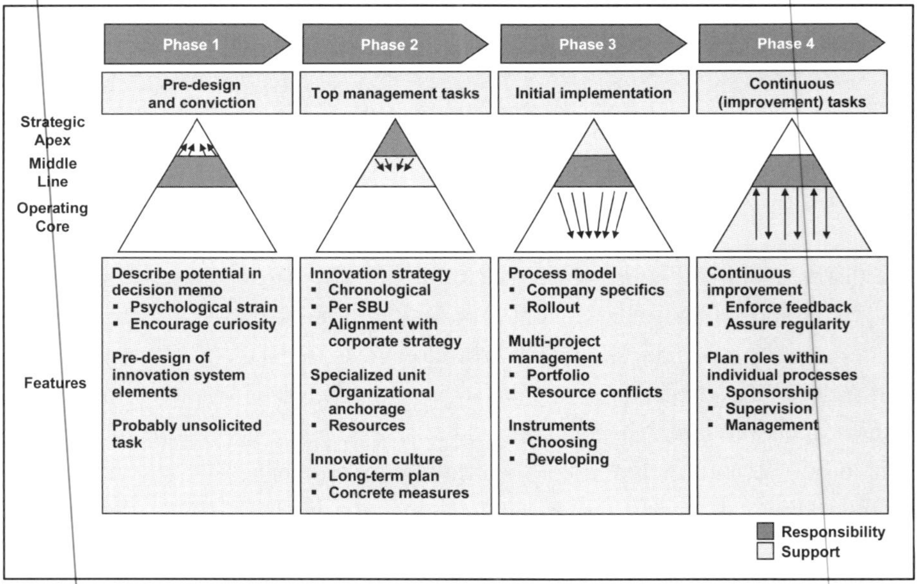

Figure 3: Pragmatic Proceeding in Four Phases

10.4 Conclusion

The implementation of an innovation management system at an LSP is a very challenging project. There are nine work packages which have to be carried out by top and middle management. The implementation process can be executed in four distinct phases. In the first, the potential has to be evaluated and an implementation plan has to be made. Once top management has been convinced, they have to influence strategy, structure and culture. Middle management can spearhead the implementation of an innovation process model, multi-project management and concrete tools. Finally, there are continuous improvement and management of individual innovation processes for which middle management, e.g., the newly created central unit, can take the lead.

We propose that future researchers supplement this conceptual analysis with empirical work such as implementation case studies.

References

Anthony, S.D.; Eyring, M.; Gibson, L. (2006). Mapping Your Innovation Strategy, *Harvard Business Review*, 84 (5), 104-113.

Burns, T.; Stalker, G. M. (1961). *The Management of Innovation*, London: Tavistock.

Hyland, P.W.; Soosay, C.A.; Sloan, T.R. (2003). Continuous Improvement and Learning in the Supply Chain, *International Journal of Physical Distribution & Logistics Management*, 33 (4), 316-336.

Soosay, C.A.; Hyland, P.W. (2005). Effect of Firm Contingencies on Continuous Innovation, *International Journal of Innovation and Technology Management*, 2 (2), 153-169.

von Zedtwitz, M. (2002). Organizational Learning through Post-Project Reviews in R&D, *R&D Management*, 32 (3), 255-268.

Zairi, M.; Al-Mashari, M. (2005). Developing a Sustainable Culture of Innovation Management: A Prescriptive Approach, *Knowledge & Process Management*, 12 (3), 190-202.

About the Editors and Authors

Stephan M. Wagner (Ph.D., University of St. Gallen) is Professor of Business Administration, holds the Kuehne Foundation Endowed Chair of Logistics Management and is Director of the Kuehne-Center for Logistics Management at WHU – Otto Beisheim School of Management, Vallendar, Germany. Prior to joining academia, he was director of SCM for a Swiss-based technology group and senior manager at a renowned consulting firm. He also taught part-time at numerous schools, including Swiss Federal Institute of Technology Zurich, Johannes Gutenberg-University Mainz and Ludwig-Maximilians-University Munich.

Stephan is author and editor of eight books and nearly hundred book chapters and articles. Previous work has appeared in journals such as *Journal of Business Logistics*, *Journal of Purchasing & Supply Management*, *International Journal of Production Research, Journal of Management, and Journal of Business Research*. He is an Associate Editor of the *Journal of Supply Chain Management* as well as the *Journal of Purchasing & Supply Management.*

Christian Busse (M.Sc., TU Berlin) is a Ph.D. student and research assistant at WHU – Otto Beisheim School of Management in Vallendar, Germany. Christian undertook a five-year joint honors program in Industrial Engineering and Management at the Technical University of Berlin and the Universidad Complutense de Madrid. In 2002, he graduated at the top of his class.

After a brief period as a market researcher in London he began to work for The Boston Consulting Group in Düsseldorf where he focused on strategic logistics and SCM projects. Here Christian became acquainted with the challenges of managing LSPs. He went back to university in 2005 to investigate innovation management at LSPs.

Les B. Artman (MBA, Case Western Reserve University) recently retired from Accenture as a partner in the firm's Supply Chain Strategy Practice in Cleveland. In nearly 30 years in the supply chain/logistics area, he spent time in industry with B.F. Goodrich Chemical and Itek Graphic Products, and then spent the bulk of his career in consulting. He has helped many manufacturers, retailers and distributors such as Dell, General Motors, Quaker, Sara Lee, Wal-Mart and Whirlpool to establish supply chains as a competitive differentiator. Les has been a long-time member of the Council of Supply Chain Management Professionals where he has been a frequent speaker at annual conferences and local roundtables.

Ronald H. Ballou (Ph.D., Ohio State University) is Professor Emeritus of Operations at Case Western Reserve University. He has published numerous articles in leading academic journals such as *Journal of Business Logistics*, *International Journal of Physical Distribution & Logistics Management*, *International Journal of Logistics Management*, *Transportation Science*, *Industrial Marketing Management*, and *Journal of Operations Management*. Additionally, Ronald is author of a number of books on logistics, including the definitive *Business Logistics/Supply Chain Management* (5th edition, Prentice Hall). He has provided advice to many companies on logistics matters, especially concerning logistics network design.

Stanton G. Cort (DBA, Harvard Business School) is Associate Professor at the Weatherhead School of Management, Case Western Reserve University. For 30 years, he has studied, written and consulted for companies in the marketing channel and distribution field. His thoughts have been published in, among others, *Journal of Marketing Research*, *Journal of Business Research*, *Journal of Retailing*, *Business Economics*, and *Harvard Business Review*. Stanton's research deals with the identification and acquisition of competitive edge through the strategic usage of information value, outsourcing or integration. Stanton teaches numerous courses on supply chain management and new product innovation.

A. Dale Flowers (DBA, Indiana University) is Associate Professor of Operations Management at the Weatherhead School of Management, Case Western Reserve University. Dale serves as Chair of the Supply Chain Management Initiative at the School, and was a founding Co-Director of the university's Institute for Management and Engineering. He possesses an extensive teaching, research and consulting background acquired over the past 30+ years. He has published articles in journals such as *Production & Operations Management*, *Journal of Operations Management*, *IIE Transactions*, *Operations Research*, and *Interfaces*. He is a Principal in both Storm Software Inc. and Supply Chain Associates.

J. Rod Franklin (Doctorate in Management, Case Western Reserve University) is a Vice President responsible for the development of complex supply chain solutions at Kuehne + Nagel, Switzerland, where he led development efforts in the areas of asset recovery services, reverse logistics activities, RFID and lead logistics services. He has extensive experience in operations, supply chain management, and information technology. Rod has held management positions at Viacore, ENTEX Information Services, Digital Equipment, Arthur Young & Company, Booz Allen Hamilton, Theodore Barry & Associates, Cameron Iron Works, and General Motors. He is a registered professional engineer, a member of the IIE, PDMA, INFORMS and SME management science and product development groups, and a frequent speaker on supply chain and product development issues.

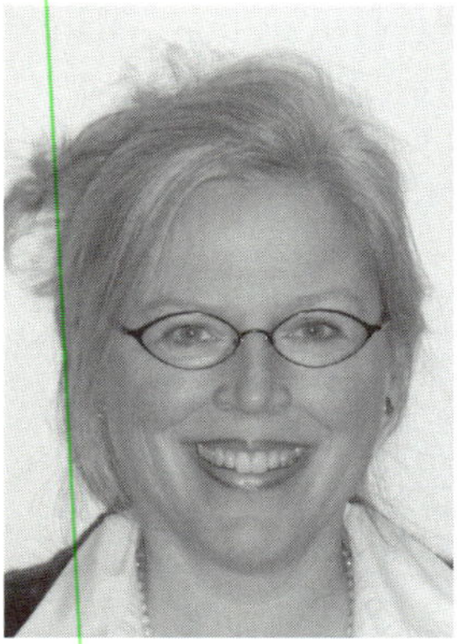

Britta Gammelgaard (Ph.D., Copenhagen Business School) is Associate Professor of Logistics and Supply Chain Management at Copenhagen Business School, where she is head of the Diploma Program in SCM. Among her main research interests are logistics innovation as well as trends and developments of the logistics industry. Britta's work has been published in international journals, such as *Journal of Business Logistics*, *International Journal of Physical Distribution & Logistics Management*, and *Transportation Journal*. She is also a member of the Editorial Board of *JBL* and Associate Editor of *Journal of Supply Chain Management*. She has been a member of the CSCMP European Conference since its inception in 2005, and in 2006 she initiated and chaired the first CSCMP European Research Seminar.

Mark Goh (Ph.D., University of Adelaide) is Director (Industry Research) at the Logistics Institute - Asia Pacific, Principal Researcher at the Centre for Transportation Research, and Associate Professor at the Department of Decision Sciences, National University of Singapore Business School. Mark has also served as Director of Supply Chain Solutions for Asia/Middle East with APL Logistics and held appointments as a Visiting Professor at Chulalongkorn University, Commonwealth Fellow, Citibank International Fellow, Visiting Research Fellow at UMIST, Visiting Scholar at Beijing University, Visiting Professor at Melbourne University, and Adjunct Professor at the University of South Australia. Some of his recent research appeared in *Production and Operations Management*, *Naval Research Logistics*, *European Journal of Operational Research*, *SIAM Journal of Optimization*, and *International Journal of Physical Distribution and Logistics Management*.

Adegoke Oke (Ph.D., Cranfield University) is Assistant Professor of Quantitative Business Analysis in the Department of Economics, Finance and Marketing at Arizona State University. He has presented his work at a number of conferences and published in the *International Journal of Operations and Production Management*, *Journal of Operations Management*, *International Journal of Production Economics*, *Journal of Change Management: An International Journal*, and others. Adegoke is an active member of professional bodies such as POMS, Decision Sciences Institute, European Operations Management Association, and the International Society of Inventories Research. Prior his academic career, Adegoke spent nearly a decade in industry as a Project Manager with Shell in Nigeria and an Operations Analyst with Allied Steel and Wire Manufacturing Company in the UK.

Robert de Souza (Ph.D., Loughborough University of Technology) is the Executive Director of The Logistics Institute – Asia Pacific, an Adjunct Professor at the School of Industrial and Systems Engineering at Georgia Institute of Technology and a Senior Fellow at the Department of Industrial and Systems Engineering at the National University of Singapore. Previously, Robert held senior management positions in industry as Executive Vice President for Asia Pacific, Corporate Senior Vice President and Global Chief Knowledge Officer of global IT-related companies and as co-founder, Vice Chairman and CEO of a Singapore start-up company up to its successful acquisition.

Thomas W. Speh (Ph.D., Michigan State University) is Professor of Marketing and James Evans Rees Distinguished Professor of Distribution at the Richard T. Farmer School of Business, Miami University, Oxford. Tom has co-authored the leading text in B-to-B marketing, *Business Marketing Management* (9th edition, Thomson Publishing Company), and written many professional and academic articles in *Journal of Marketing*, *Journal of Retailing*, *Journal of the Academy of Marketing Science*, *Journal of Business-to-Business Marketing*, *Journal of Business Logistics*, *Harvard Business Review*, *MIT Sloan Management Review*, and many other journals. He is past president of both the Council of Logistics Management (now the Council of Supply Chain Management Professionals) and the Warehousing Education and Research Council. Tom is a frequent speaker at logistics and supply chain conferences around the world.

Zach G. Zacharia (Ph.D., University of Tennessee) is an Assistant Professor of Supply Chain Management in the Department of Management at Lehigh University. Zach has published in *Journal of Business Logistics*, *Journal of Retailing*, *International Journal of Vehicle Design*, and *Transportation Research Record*, and written several chapters in the book *Supply Chain Management* (John T. Mentzer, ed.). Zach specializes in new product development and the role of logistics, as well as in the theory and implementation of inter-firm relationships in supply chains. He has won numerous academic awards, including the 2002, 2003, 2004 and 2005 Research and Creative Activities Grant by Texas Christian University. Zach is the committee chair for the Council of Supply Chain Management Professionals (CSCMP) Academic Case Studies program and is President of the CSCMP roundtable in Dallas, Texas.

Christian Martin Meyer

Integration des Komplexitätsmanagements in den strategischen Führungsprozess der Logistik

Schriftenreihe Logistik der Kühne-Stiftung. Band 12
2007. XVI + 206 Seiten, 43 Abbildungen, 11 Tabellen, kartoniert
CHF 42.– / EUR 27.50
ISBN 978-3-258-07197-8

Seit Jahren verändern sich die Wertschöpfungsketten in ihren Strukturen und Prozessen rasch. Megatrends wie die Globalisierung, die Verkürzung der Produktlebezyklen und die zunehmende Kundenorientierung steigern in der Querschnittfunktion Logistik die Komplexität. Diese muss deshalb zielgerichtet bewältigt werden und als Gestaltungsdimension in den Führungsprozess der Logistik integriert werden.
Christian Martin Meyer entwickelt eine praktisch anwendbare Methode, die sowohl die Definition strategischer Ziele des Komplexitätsmanagements als auch die Umsetzung in der Logistik unterstützt. Basis der modular aufgebauten Methode ist ein systemtheoretisches Wirkmodell der Komplexität, das die Wirkbeziehungen zwischen Treibern und Auswirkungen der Komplexität sichtbar macht.
Das Buch vermittelt praxisorientiert die wesentlichen Grundlagen, wertvolle Detailinformationen sowie eine Vielzahl von Anregungen zum Umgang mit Komplexität in der Logistik. Es richtet sich daher sowohl an Dozenten und Studenten im akademischen Umfeld als auch an Fach- und Führungskräfte in der Logistik.

: Haupt **Haupt Verlag** Bern • Stuttgart • Wien
verlag@haupt.ch • www.haupt.ch